Industrial Organization and Trade in the Food Industries

Industrial Organization and Trade in the Food Industries

EDITED BY

Ian M. Sheldon
and Philip C. Abbott

Routledge
Taylor & Francis Group

LONDON AND NEW YORK

First published 1996 by Westview Press, Inc.

Published 2018 by Routledge
52 Vanderbilt Avenue, New York, NY 10017
2 Park Square, Milton Park, Abingdon, Oxon OX14 4RN

Routledge is an imprint of the Taylor & Francis Group, an informa business

A CIP catalog record for this book is available from the Library of Congress.

ISBN 13: 978-0-367-01204-5 (hbk)
ISBN 13: 978-0-367-16191-0 (pbk)

Contents

Preface

In April, 1993, the North Central Regional Research Project NC-194: "The Organization and Performance of World Food Systems: Implications for U.S. Food Policies" sponsored a conference called Empirical Studies of Industrial Organization and Trade in the Food Industries in Indianapolis, IN. Papers presented at that conference examined the extent of and tests for market power exercised by firms in international markets, the impact of market power on international competitiveness, the role of multinational food firms and intra-firm trade, and evaluation of trade and industrial policy under imperfect competition. This conference followed an earlier NC-194 conference in April, 1991, where the relevance of the theoretical methods in these two areas to food industry trade was explored. The Indianapolis conference placed emphasis on empirical applications of new methods linking industrial organization and trade theory that addressed important trade and policy issues for the U.S. food industries. Sixteen papers covering a variety of topics in these areas were presented at the conference and were used to develop the contents of this book.

NC-194 is a Cooperative Research Project sponsored by Agricultural Experiment Stations of the North Central Region, with cooperation from agricultural experiment stations of several states outside the region. The project also includes as participants members of foreign universities as well as researchers from several agencies of the U.S. Department of Agriculture. Funding was provided in part by a special agricultural research appropriation from the U.S. Congress. Executive offices of NC-194 are maintained at The Ohio State University.

The purpose of the sponsoring organization, NC-194, is to investigate competition, coordination, and economic performance in international markets for food and agricultural products. In contrast to previous research on international trade in agricultural products, this project focuses on (1) international marketing of high-value, value-added, processed, and semi-processed products; (2) the organization, structure, and behavior of agricultural processing and distribution firms in international markets; and (3) factors other than national differences in costs of production as determinants of world market shares and of international competitiveness.

This project has utilized analytical approaches derived from international trade theory, industrial organizational theory, marketing, organizational behavior, and political science to achieve its objectives. The underlying economic foundation is a newly evolving hybrid derived principally from the theories of industrial organization and international trade. The variety of methods in these areas is reflected in the variety of approaches utilized in the chapters of this book.

The organizing committee for the Indianapolis conference was chaired by Ian Sheldon and Philip Abbott. Other members of the organizing committee included: Azzedine Azzam, John Connor, Alan Love, Les Myers, and Emilio Pagoulatos. Support for this book and the conference were provided by the Purdue University Research Cluster of NC-194. Thanks are due to all members of the organizing committee and the authors as well as to Judy Conner for her tireless efforts in preparing this book for publication.

Ian M. Sheldon
Philip C. Abbott

1

Introduction

Ian M. Sheldon and Philip C. Abbott

While the structure and behavior of the domestic food processing and retailing system has been subject to detailed analysis over the past two decades (see Connor *et al.* 1985; Marion 1986), until recently, economic analysis of international markets for processed agricultural products, manufactured foods, and other high-value products has received very little attention in the agricultural economics literature. Most economic analysis of international markets has focused on basic agricultural commodities, which accounted for about 45 percent of the value of total U.S. food and agricultural trade in 1992 (MacDonald and Lee 1994). World food trade has become increasingly dominated by the manufactured foods sector, however. Between 1972 and 1990, the value of trade in manufactured food products grew by 442 percent, compared to growth in the value of bulk commodity trade of 337 percent, such that manufactured food trade now accounts for 64 percent of world food trade (United Nations 1990). Trade in manufactured foods is also highly concentrated among a few countries, with 24 countries accounting for 80 percent of shipments in 1990 (United Nations 1990). For the U.S., the third largest supplier of manufactured foods, exports of manufactured foods accounted for about 55 percent of total U.S. food trade in 1992 (MacDonald and Lee 1994), and for some European countries, the share of manufactured foods in total food trade is even greater, equaling 68, 74, 82 and 81 percent respectively in France, the Netherlands, Germany, and the UK (United Nations 1990).

Given the growth in world trade in manufactured foods and high-value agricultural products, it is important to recognize that industries in this sector may not fit the assumptions of the standard, perfectly competitive model. As shown by Connor *et al.* (1985) and Sutton (1991), the food processing industries in developed countries tend to have imperfectly competitive market structures characterized by high seller concentration, some degree of plant level economies of scale, and product differentiation. For example, Sutton (1991) reports that the cross-industry, cross-country four-firm sales concentration ratios for the period 1987-88 were, on average, 65 percent for the food processing sectors in the U.S. and European Community (EC). Generally, the food processing industries in both the U.S. and EC are dominated by small numbers of firms, suggesting that there is potential for oligopolistic interaction.

In addition to these characteristics, the food industry in developed countries has other key features in terms of the structure of international trade, and the types of marketing arrangements between firms. The concurrent importation and exportation of similar goods, intra-industry trade (IIT), has been documented for the food manufacturing sector by several researchers (McCorriston and Sheldon 1991; Christodoulou 1992; Hirschberg, Sheldon, and Dayton 1994). This type of trade is normally rationalized by appeal to factors such as imperfect competition, economies of scale, and product differentiation (Helpman and Krugman 1985). Also, food manufacturing firms, both in the U.S. and elsewhere, have a strong propensity for accessing foreign markets through strategies other than "arm's length" exporting, notably foreign direct investment (FDI) and branded product licensing. For example, Handy and Henderson (1994) note that in 1989, U.S. multinational corporations held at least 10 percent equity in 720 food manufacturing affiliates, and in 1990 sales from these foreign affiliates stood at $75 billion, growing at an annual rate of 11 percent over the 1980s. Also, Henderson and Sheldon (1992) have shown that the international licensing of branded food products is a frequently used marketing strategy by food manufacturing firms.

Analysis of trade in the agricultural sector has generally rested on the traditional theory of comparative advantage, which assumes perfectly competitive markets where the goods sold are homogeneous and produced under a technology of constant returns to scale. In addition, trade is assumed to take the form of arm's length exports and imports, where the structure of trade is inter-industry in nature. Virtually all of the empirical research on the impact of trade liberalization due to the recent GATT round was based on this theoretical framework (see Tyers and Anderson 1992). However, the characteristics of the food processing sector suggest that this model may not be especially relevant for understanding the nature of trade and international competition in manufactured food products.

In an earlier volume, the possible methodologies were laid out for analyzing trade and industrial organization in the food industries (Sheldon and Henderson 1992). Empirical analysis of such issues is the core of this book.

Methodologies

Much of the research contained in the following chapters draws on theoretical and empirical developments that have taken place in the industrial organization and international economics literatures over the past two decades. While the intention here is not to present a detailed survey of these developments, it is useful to highlight some of the salient characteristics of this work in order to provide a context for the remainder of the book.

In many ways, the research reported in this book represents an effort to integrate industrial organization analysis with that of international economics, and to apply this "theoretical hybrid" to the food and agricultural sector. However, while one can now usefully talk about the crossover of ideas between industrial organization and international economics, the evolution of such a merger of ideas really came out of each branch independently adapting aspects from the other. Specifically, in the early 1970s, industrial organization economists working within the structure-conduct-performance (SCP) paradigm began to introduce international variables into cross-industry regression studies, in the belief that such variables would affect competition in domestic markets. At about the same time, international trade economists, who until then had worked entirely within the Heckscher-Ohlin-Samuelson (HOS) tradition, began to borrow ideas from industrial organization by allowing for firms selling differentiated products in international markets produced under a technology of increasing returns. Consequently, it is useful to focus separately on the theoretical and empirical methodologies of these two literatures.

Industrial Organization and Trade

In the 1970s, most research in industrial organization followed the SCP approach, originally pioneered by Bain (1951). Although based largely on informal theorizing, this approach hypothesized that market structure (seller concentration, barriers to entry, etc.) affects the market conduct of firms (pricing, advertising, etc.) which, in turn, results in market performance as measured by indices such as profits, efficiency, and rates of innovation. This framework generated a large number of empirical studies based on simple, reduced-form regression analyses of cross-industry data sets, the food manufacturing sector included (see Connor *et al.* 1985). Generally, these studies can be characterized as seeking a causal relationship between some measure of industry profits and a series of variables proxying market structure. Schmalensee (1989) has exhaustively surveyed this body of research, and while it has been subject to some well-merited criticism (Pagoulatos 1992; Perloff 1992), he concludes:

...inter-industry research in industrial organization should generally be viewed as a search for empirical regularities, not as a set of exercises in structural estimation....research in this tradition has indeed uncovered many stable, robust, empirical regularities. (p. 1000)

In terms of the research presented in this book, a key feature of empirical analysis of the SCP framework prior to the 1970s was the implicit assumption that industries were operating under autarky. However, as international trade expanded in the post-war period, industrial organization economists recognized that increasing import competition in domestic markets might have an important disciplining effect on firms' behavior. As a result, many of the later empirical studies using the SCP framework incorporated international variables such as import to domestic sales ratios as a factor affecting the exercise of domestic market power. The theoretical logic behind this has been laid out in a number of papers (Lyons 1981; Jacquemin 1982; Sugden 1983). The basic argument is that industry price-cost margins will vary inversely with the share of imports, and also, the effect of import competition on price-cost margins will interact with both seller concentration and the supply elasticity of imports. Specifically, as Caves (1985) states, import competition has to be able to discipline and to have something to discipline. A survey by Lyons (1981) of 23 cross-industry regression studies supports the hypothesis that imports can restrict the ability of firms to raise prices above marginal cost.

In contrast to the role of imports, the impact of exports on firms' domestic market behavior and performance is less clear. On the one hand, if a small country assumption holds, and domestic firms are unable to price discriminate between home and foreign markets, exporting firms will be constrained to behaving competitively in both domestic and foreign markets. Alternatively, if a domestic monopolist is able to price discriminate, the opportunity to export can result in an increase in domestic price-cost margins. However, as Pugel (1980) has shown, this relationship is not unambiguous since overall price-cost margins for the exporting firm(s) are a weighted average of the margin on domestic and foreign sales. Even if profits increase through discrimination, price-cost margins may not, as increases in the domestic margin may be offset by decreases in the export margin. To get definite predictions here, account has to be taken of the exporters' cost conditions (Huveneers 1981). Perhaps not surprisingly, the empirical work on exports and domestic market performance has not produced uniform results. Studies by

Pagoulatos and Sorenson (1976), and Neuman, Böbel and Haid (1979) show that exports reduce industrial profitability, while Geroski (1982) finds the opposite.

In the last fifteen years, industrial organization has undergone somewhat of a transformation from the largely empirical tradition of the SCP approach to a more theoretically rigorous methodology, characterized by the ubiquitous use of game-theoretic methods, and is sometimes described as the "new industrial organization" (NIO) (see Jacquemin 1987; Pagoulatos 1992). Tirole's (1989) widely used textbook characterizes the modern approach to industrial organization, and much of this analysis has formed the basis for recent developments in international economic theory.

In terms of empirical work, the NIO has seen a parallel development in empirical methods in industrial organization, usually referred to as the "new empirical industrial organization" (NEIO). As Pagoulatos (1992) notes, the new empirical methodology has largely developed as a result of dissatisfaction with the SCP approach to analyzing market power. Survey articles by Bresnahan (1989) and Perloff (1992) show that, in the last decade, relatively complete structural econometric models based on formal profit-maximizing theories have been used to estimate the degree of market power in specific industries using time-series data. To date, there have been in excess of ten applications of the new methodology to specific industries in the food manufacturing and related sectors, including Just and Chern (1980), Lopez (1984), Schroeter (1988) and Azzam and Pagoulatos (1990). However, very few studies have used this type of analysis with respect to export markets, the exceptions being Buschena and Perloff (1991), Karp and Perloff (1989, 1993), and Lopez and You (1993). In addition, econometric analysis of Krugman's (1987) pricing to market hypothesis (PTM), whereby exporters undertake price discrimination across export markets with respect to exchange rate movements, has only been conducted on the basis of a static monopoly profit-maximization problem (Knetter 1989; Pick and Park 1991).

Trade Theory and Industrial Organization

It might be argued that the incorporation of industrial organization concepts into trade theory, the so-called "new trade theories" (NTTs), has been far more radical than the recognition of international trade by industrial organization economists. To quote Krugman (1989):

> Traditional theory was, by the late 1970s, a powerful monolithic structure in which all issues were analyzed using variants of a single model. The new literature has successfully broken the grip of that single approach.

Increasingly, international economics, like industrial organization, is becoming a field where many models are taught and research is an eclectic mixture of approaches. (p. 1214)

Traditional trade theory, embodied in the Heckscher-Ohlin-Samuelson (HOS) framework, was largely based on general-equilibrium, competitive models of the international economy (Ethier 1994). In addition, international economists paid limited attention to empirical testing, Leamer's (1984) work being an obvious exception. Unfortunately, as pointed out by many observers, the predictions of the neoclassical model do not fit well the observed facts of international trade. In particular, neither the existence of intra-industry trade (IIT) nor the prevalence of multinational firms sits very comfortably with the HOS framework (Ethier 1994).

Intra-Industry Trade and Multinationals

IIT has become an increasingly important phenomenon in international trade in the post-war period. It was first observed in empirical work on the evolution of the EC by Verdoorn (1960) and Balassa (1965). Since then an extensive literature has shown broad evidence for such trade in both the developed and less developed countries (see Greenaway and Milner 1986a). Conventional trade theory, which predicts inter-industry trade on the basis of comparative advantage, cannot rationalize the existence of IIT. In recent years, a large theoretical literature has emerged that attempts to explain IIT (Helpman and Krugman 1985). These theoretical developments have predominantly emphasized the role of imperfect market structures, scale economies, and product differentiation as the basic factors determining IIT. Probably the best known and most general models are based on a structure of monopolistic competition, the major contributions having been synthesized by Helpman and Krugman (1985), drawing on earlier work by Krugman (1979, 1980, 1981), Lancaster (1980) and Helpman (1981).

There have been several econometric studies of the determinants of IIT (see Greenaway and Milner 1986a). Most of the empirical work has set out to estimate a simple, reduced-form regression model where the dependent variable is an index of IIT for industry i at time t, and the explanatory variables are a vector of either industry and/or country characteristics based on the theory outlined. Similar to the empirical work on the SCP approach, most studies estimate this type of regression over a cross-section of industries, using either bilateral or multilateral trade data, while some studies, notably Balassa and Bauwens (1987), have adopted a multi-industry, multi-country framework. As researchers in this field acknowledge (Greenaway and Milner 1986b), because there is no completely general model of IIT, the econometric analysis of IIT is difficult both from a methodological and a practical viewpoint. Consequently, most researchers have resorted to:

> ...'identifying' (using regression techniques) the cross-sectional characteristics of their data sets, i.e. to describing a range of sources of influence on IIT that are suggested by, or consistent with, the various models of IIT. (Greenaway and Milner 1986b, p. 5)

Notwithstanding these problems, the results of this econometric analysis indicate fairly robust and consistent support for the theory as laid out earlier. Proxy variables for market structure, product differentiation and economies of scale do have some explanatory power along with other control variables such as tariffs (see Greenaway and Milner 1986a).

In the case of multinational firms and FDI, as noted by Ethier (1994), conventional trade theory has "nothing to say" (p. 109). As a result of the assumptions about perfect competition and constant returns to scale, it is perhaps not unexpected that neoclassical trade theory does not deal with multinationals, and, in addition, FDI is explicitly ruled out by the traditional model. However, as Ethier (1994) notes, even if FDI is introduced into the HOS framework, the predictions that FDI will be driven by differences in factor endowments and negatively correlated with direct exports are contradicted by the stylized facts, since a large amount of FDI occurs between developed countries.

Apart from a small number of papers by, for example, Helpman (1984), Markusen (1984), and Ethier (1986), which incorporate multinational firms into a general equilibrium framework characterized by imperfect competition, economies of scale and product differentiation, trade theory has not really developed a general theory of multinationals and FDI. However, as Ethier (1986, 1994) points out, an extensive, informal, microeconomics literature on multinationals has evolved, drawing on ideas from industrial organization, transactions cost theory and business economics (see Caves 1982 for a survey). Along with this there is a small but growing empirical literature that uses methods similar to those used in industrial organization research for testing explanations of FDI (see Buckley 1985 and Casson 1987).

Dunning (1981) has described this approach to explaining the existence of multinational firms as the "OLI framework", where OLI stands for the specific advantages of ownership, location, and internalization. *Ownership advantages* refer to a firm's particular assets that are not owned by other firms, which it wants to both protect and receive a return on. In the case of a multinational firm, these advantages might relate to firm-specific assets such as patents, trademarks, brand loyalty, research and development resources, managerial skills and other "headquarters' services". *Location advantages* most often relate to the benefits that a multinational corporation (MNC) reaps from avoiding various impediments to direct trade, hence the concept of "tariff-jumping". *Internalization advantages* is probably the major reason forwarded for firms to engage in FDI. Internalization is a situation where intermediate product markets are integrated within a multinational firm. Intermediate products are either tangible, semi-processed or intangible assets, such as patents, trademarks and human capital. While such assets may be transferred physically at little expense, transferring them at arms' length through the market may incur transactions costs or may be subject to market imperfections. These costs can be avoided if the firm creates an internal market for the transfer of the asset. Increasingly, empirical work is attempting to incorporate such factors into analysis of FDI.

Trade Policy and Industrial Organization

While the NTTs that deal with the positive economic analysis of aspects such as trade structure are relatively non-controversial, the same cannot be said for the NTTs that deal with trade policy under conditions of imperfect competition. As Helpman and Krugman (1989) note:

> Although the positive theory of trade under imperfect competition has now reached a certain maturity and acceptance, the same cannot be said of the theory of trade policy under imperfect competition. (p. 2)

Most controversially, it has been shown by Brander and Spencer (1985) that the existence of oligopolistic markets may provide a rationale for activist trade policies such as export subsidies (see Krishna and Thursby 1990 for a review of this literature). Although such "rent-shifting" arguments have been shown to be a rather special case (Dixit 1987), the NTTs have made a significant contribution to understanding how imperfect competition can affect the gains from trade liberalization (Cox and Harris 1985 and Smith and Venables 1988), and how different trade instruments can have differential welfare effects when markets are oligopolistic, (e.g. Harris 1985 and Krishna 1989).

Empirical analysis in this area is pretty much in its infancy, and the methods employed are not particularly sophisticated. In terms of the rent-shifting models, virtually all empirical research has been based on some form of partial equilibrium simulation model for a particular industry (Baldwin and Krugman 1988; Dixit 1988; Smith and Venables

1988). Research on the benefits of trade liberalization under imperfect competition has tended to use computable general equilibrium models (Cox and Harris 1985; Hertel and Laclos 1994; Kehoe 1994), but few of these studies have explicitly analyzed the food industries.

These approaches have been thoroughly reviewed in papers by Richardson (1990) and Sheldon (1992), for example. The basic method is to specify a theoretical model that captures certain features of imperfectly competitive markets, such as oligopolistic interaction, product differentiation, and scale economies. Each model contains a number of parameters and endogenous variables, such as prices and quantities. Some of the parameters are taken from external estimates, while the rest are calibrated to the model in order to reproduce the chosen base-period data. Once these models are calibrated they can be used for simulation, what Richardson (1990) has called "counterfactual" exercises. Essentially, the models described are maintained as true, and generate the observed data for the base-line period. The counterfactual step is to arbitrarily alter one part of the model, assuming the other parameters and/or variables remain constant. The new equilibrium is then calculated and compared with the base-line equilibrium. Hence, the aim is not to test the validity of the underlying model, but to gain an idea of the broad effects of trade policy, assuming such a market structure exists.

Most of the empirical analysis of rent-shifting indicates that the welfare benefits of trade interventions tend to be rather small (Dixit 1988), while the empirical analysis of trade liberalization indicates that specialization in the presence of economies of scale generates production gains, the opening up of imperfectly competitive markets to imports yields competitive gains, and trade widens consumer choice. For example, Cox and Harris (1985) have shown in their evaluation of the U.S./Canada free trade agreement that the presence of economies of scale generates considerably larger welfare gains than those predicted by neoclassical trade theory. Similar analysis by Smith and Venables (1988) of the move to greater economic integration in the EC indicates that in a framework of product differentiation and economies of scale, the gains from removing trade barriers are much larger than those normally associated with conventional customs union analysis.

Methodologies and This Book

The theoretical and empirical methodologies that characterize research in industrial organization and international economics are rich and diverse. The common characteristic of this research has been the attempt to capture aspects such as imperfect competition, economies of scale, and product differentiation in empirical analysis of international markets. Given that such factors seem fairly typical of trade and market structure in the food manufacturing industries, it is perhaps not surprising that they are a common thread running through the succeeding chapters in this book.

The preceding summary of methodologies used in industrial organization and international economics research also indicates a variety of empirical techniques and theoretical models which can be utilized to conduct research in this area. This is also a characteristic of the research presented in the book. Many of the chapters are based on the cross-sectional econometric methodology adopted both in the industrial organization literature and the empirical research on intra-industry trade and foreign direct investment. They focus on topics ranging from the interaction between foreign trade, market structure, business cycles, and profit margins through the relationship between levels of protection and political lobbying activities in the food manufacturing sector, to cross-industry, cross-country analysis of intra-industry trade. In addition, several chapters, which focus on the pricing to market hypothesis and also foreign direct investment, utilize methods associated with the NEIO, and even go some way toward synthesizing these methods with the earlier techniques used by industrial organization economists. Several chapters also utilize the theoretical and empirical methodologies of the NTTs to analyze a series of issues ranging from the analysis of product differentiation in the sweetener market under conditions of imperfect competition, through analysis of competitiveness and technological change using a monopolistic competition-type model, to the differential effects of quotas and tariffs when markets are oligopolistic.

Industrial organization and international economics have undergone important paradigm shifts in recent years in terms of both theory and empirical methods. Much of this shift was due to a recognition that markets are becoming more international in focus, and because the traditional assumptions of perfect competition, constant returns to scale, and homogeneous goods are simply violated by the facts. With the recognition that international trade in manufactured food does not fit the traditional model, the chapters in this book represent an attempt to apply, synthesize and develop the findings of the industrial organization-international trade literature with respect to this sector.

References

Azzam, A.M. and E. Pagoulatos. 1990. "Testing Oligopolistic and Oligopsonistic Behavior: An Application to the US Meat Packing Industry." *Journal of Agricultural Economics* 41:362-369.

Bain, J.S. 1951. "Relation of Profit Rate to Industry Concentration: American Manufacturing, 1936-1940." *Quarterly Journal of Economics* 65:293-324.

Balassa, B. 1965. *Economic Development and Integration.* Mexico: Centro de Estudios Monetarios Latinoamericanos.

Balassa, B. and L. Bauwens. 1987. "Intra-Industry Specialization in a Multi-Country and Multi-Industry Framework." *Economic Journal* 97:923-939.

Baldwin, R. and P.R. Krugman. 1988. "Market Access and International Competition: A Simulation Study of 16K Random Access Memories," in R.C. Feenstra, ed., *Empirical Methods for International Trade.* Cambridge, MA: MIT Press.

Brander, J.A. and B.J. Spencer. 1985. "Export Subsidies and International Market Share Rivalry." *Journal of International Economics* 18:83-100.

Bresnahan, T.F. 1989. "Empirical Studies of Industries With Market Power", in R. Schmalensee, R. and R. Willig, eds., *Handbook of Industrial Organization.* Amsterdam: North Holland.

Buckley, P.J. 1985. "Testing Theories of the Multinational Enterprise: A Review of the Evidence," in P.J. Buckley and M. Casson, eds., *The Economic Theory of the Multinational Enterprise.* New York, NY: St. Martin's Press.

Buschena, D.E. and J.M. Perloff. 1991. "The Creation of Dominant Firm Market Power in the Coconut Oil Export Market." *American Journal of Agricultural Economics* 73:1000-1008.

Casson, M. 1987. "Multinational Firms," in R. Clarke and T. McGuinness, eds. *The Economics of the Firm.* Oxford: Basil Blackwell.

Caves, R.E. 1982. *Multinational Enterprise and Economic Analysis.* Cambridge: Cambridge University Press.

_____. 1985. "International Trade and Industrial Organization: Problems, Solved and Unsolved." *European Economic Review* 28:377-395.

Christodoulou, M. 1992. "Intra-Industry Trade in Agrofood Sectors: The Case of the EEC Meat Market." *Applied Economics* 24:875-884.

Connor, J.M., R.T. Rogers, B.W. Marion, and W.F. Mueller. 1985. *The Food Manufacturing Industries: Structures, Strategies, Performance, and Policies.* Lexington, MA: Lexington.

Cox, D. and R. Harris. 1985. "Trade Liberalization and Industrial Organization: Some Estimates for Canada." *Journal of Political Economy* 93:115-145.

Dixit, A.K. 1987. "Strategic Aspects of Trade Policy," in T.F. Bewley, ed., *Advances in Economic Theory.* Cambridge: Cambridge University Press.

_____. 1988. "Optimal Trade and Industrial Policy for the U.S. Automobile Industry," in R.C. Feenstra, ed., *Empirical Methods for International Trade.* Cambridge, MA: MIT Press.

Dunning, J.H. 1981. "Explaining the International Direct Investment Position of Countries: Towards a Dynamic or Developmental Approach." *Weltwirtschaftliches Archiv* 117:30-64.

Ethier, W.J. 1986. "The Multinational Firm." *Quarterly Journal of Economics* 102:805-833.

Ethier, W.J. 1994. "Conceptual Foundations from Trade, Multinational Firms, and Foreign Direct Investment Theory," in M.E. Bredahl, P.C. Abbott, and M.R. Reed, eds., *Competitiveness in International Food Markets*. Boulder, CO: Westview Press.

Geroski, P. 1982. "Simultaneous Equations Models of the Structure-Performance Paradigm." *European Economic Review* 19:147-158.

Greenaway, D. and C. Milner. 1986a. *The Economics of Intra-Industry Trade*. Oxford: Basil Blackwell.

_____. 1986b. "Growth and Significance of Intra-Industry Trade." Paper presented at the International Economics Study Group.

Handy, C.R. and D.R. Henderson. 1994. "Assessing the Role of Foreign Direct Investment in the Food Manufacturing Industry," in M.E. Bredahl, P.C. Abbott, M.R. Reed, eds., *Competitiveness in International Food Markets*. Boulder, CO: Westview Press.

Harris, R. 1985. "Why Voluntary Export Restraints are 'Voluntary'." *Canadian Journal of Economics* 18: 799-801.

Helpman, E. 1981. "International Trade in the Presence of Product Differentiation, Economies of Scale, and Monopolistic Competition." *Journal of International Economics* 11:305-340.

_____. 1984. "A Simple Theory of International Trade with Multinational Corporations." *Journal of Political Economy* 92:451-471.

Helpman, E. and P.R. Krugman. 1985. *Market Structure and Foreign Trade*. Cambridge, MA: MIT Press.

_____. 1989. *Trade Policy and Market Structure*. Cambridge, MA: MIT Press.

Henderson, D.R. and I.M. Sheldon. 1992. "International Licensing of Branded Food Products." *Agribusiness: An International Journal* 8:399-412.

Hertel, T.W. and K. Laclos. 1994. "Trade Policy Reform in the Presence of Product Differentiation and Imperfect Competition: Implications for Food Processing Activity," in M. Hartmann, P.M. Schmitz, and H. Von Witzke, eds., *Agricultural Trade and Economic Integration in Europe and in North America*. Kiel: Wissenschaftsverlag Vauk Kiel KG.

Hirschberg, J.G., I.M. Sheldon, and J.R. Dayton. 1994. "An Analysis of Bilateral Intra-Industry Trade in the Food Processing Sector." *Applied Economics* 26:159-167.

Huveneers, C. 1981. "Price Formation and the Scope for Oligopolistic Conduct in a Small Open Economy." *Recherches Economiques de Louvain* 47:209-242.

Jacquemin, A. 1982. "Imperfect Market Structure and International Trade-Some Recent Research." *Kyklos* 35:75-93.

_____. 1987. *The New Industrial Organization: Market Forces and Strategic Behavior*. Cambridge, MA: MIT Press.

Just, R.E. and W.S. Chern. 1980. "Tomatoes, Technology, and Oligopsony." *Bell Journal of Economics* 11:584-602.

Karp, L. and J.M. Perloff. 1989. "Oligopoly in the Rice Export Market." *Review of Economics and Statistics* 71:462-470.

_____. 1993. "A Dynamic Model of Oligopoly in the Coffee Export Market." *American Journal of Agricultural Economics* 75:448-457.

Kehoe, T.J. 1994. "Assessing the Impact of North American Free Trade", in M. Hartmann, P.M. Schmitz, and H. Von Witzke, eds., *Agricultural Trade and Economic Integration in Europe and in North America.* Kiel: Wissenschaftsverlag Vauk Kiel KG.

Knetter, M.M. 1989. "Price Discrimination by US and German Exporters." *American Economic Review* 79:198-210.

Krishna, K. 1989. "Trade Restrictions as Facilitating Practices." *Journal of International Economics* 26: 251-270.

Krishna, K. and M. Thursby. 1990. "Trade Policy with Imperfect Competition: A Selective Survey", in C. Carter, A. McCalla, and J. Sharples, eds., *Imperfect Competition and Political Economy: The New Trade Theory in Agricultural Trade Research.* Boulder, CO: Westview Press.

Krugman, P.R. 1979. "Increasing Returns, Monopolistic Competition, and International Trade." *Journal of International Economics* 9:469-480.

_____. 1980. "Scale Economies, Product Differentiation and the Pattern of Trade." *American Economic Review* 70:950-959.

_____. 1981. "Intra-Industry Specialization and the Gains from Trade." *Journal of Political Economy* 89:959-973.

_____. 1987. "Pricing to Market When the Exchange Rate Changes", in S.W. Arndt and J.D. Richardson, eds., *Real-Financial Linkages Among Open Economies.* Cambridge, MA: MIT Press.

_____. 1989. "Industrial Organization and International Trade," in R. Schmalensee and R.D. Willig, eds., *Handbook of Industrial Organization.* Amsterdam: North-Holland.

Lancaster, K. 1980. "Intra-Industry Trade under Perfect Monopolistic Competition." *Journal of International Economics* 10:151-176.

Leamer, E.E. 1984. *Sources for International Comparative Advantage: Theory and Evidence.* Cambridge, MA: MIT Press.

Lopez, R.A. and Z. You. 1993. "Determinants of Oligopsony Power: The Haitian Coffee Case." *Journal of Development Economics* 41:275-284.

Lopez, R.E. 1984. "Measuring Oligopoly Power and Production Responses of the Canadian Food Processing Industry." *Journal of Agricultural Economics* 35:219-230.

Lyons, B. 1981. "Price-Cost Margins, Market Structure, and International Trade," in D. Currie, D. Peel, and W. Peters, eds., *Microeconomic Analysis: Essays in Microeconomics and Economic Development.* London: Croom Helm.

MacDonald, S. and J.E. Lee. 1994. "Assessing the International Competitiveness of the United States Food Sector," in M.E. Bredahl, P.C. Abbott, and M.R. Reed, eds., *Competitiveness in International Food Markets.* Boulder, CO: Westview Press.

Marion, B.W. 1986. *The Organization and Performance of the U.S. Food System.* Lexington, MA: Lexington Books.

Markusen, J.R. 1984. "Multi-Plant Economies, and the Gains from Trade." *Journal of International Economics* 16:205-226.

McCorriston, S. and I.M. Sheldon. 1991. "Intra-Industry Trade and Specialization in Processed Agricultural Products: The Case of the U.S. and the E.C." *Review of Agricultural Economics* 13:173-184.

Neumann, M., I. Böbel, and A. Haid. 1979. "Profitability, Risk and Market Structure in West German Industries." *Journal of Industrial Economics* 27:227-242.

Pagoulatos, E. 1992. "Empirical Studies of Industrial Organization and Trade: A Selective Survey," in I.M. Sheldon and D.R. Henderson, eds., *Industrial Organization and International Trade: Methodological Foundations for International Food and Agricultural Market Research*. NC-194 Research Monograph No. 1. Columbus, OH: Ohio State University.

Pagoulatos, E. and R. Sorenson. 1976. "Domestic Market Structure and International Trade: An Empirical Analysis." *Quarterly Review of Economics and Business* 16:45-59.

Perloff, J.M. 1992. "Econometric Analysis of Imperfect Competition and Implications for Trade Research," in I.M. Sheldon and D.R. Henderson, eds., *Industrial Organization and International Trade: Methodological Foundations for International Food and Agricultural Market Research*. NC-194 Research Monograph No.1. Columbus, OH: Ohio State University.

Pick, D.H. and T.A. Park. 1991. "The Competitive Structure of US Agricultural Exports." *American Journal of Agricultural Economics* 73:133-141.

Pugel, T.A. 1980. "Foreign Trade and U.S. Market Performance." *Journal of Industrial Economics* 29:119-130.

Richardson J.D. 1990. "International Trade, National Welfare, and the Workability of Competition: A Survey of Empirical Estimates," in C.A.Carter, A.F.McCalla and J.A. Sharples, eds., *Imperfect Competition and Political Economy-The New Trade Theory in Agricultural Trade Research*. Boulder, CO: Westview Press.

Schmalensee, R. 1989. "Inter-Industry Studies of Structure and Performance," in R. Schmalensee and R. Willig, eds., *Handbook of Industrial Organization*. Amsterdam: North Holland.

Schroeter, J.R. 1988. "Estimating the Degree of Market Power in the Beef Packing Industry." *Review of Economics and Statistics* 70:158-162.

Sheldon, I.M. 1992. "Imperfect Competition and International Trade: The Use of Simulation Techniques," in I.M. Sheldon and D.R. Henderson, eds., *Industrial Organization and International Trade: Methodological Foundations for International Food and Agricultural Market Research*. NC-194 Research Monograph No.1. Columbus, OH: Ohio State University.

Sheldon, I.M. and D.R. Henderson, eds. 1992. *Industrial Organization and International Trade: Methodological Foundations for International Food and Agricultural Market Research*, NC-194 Research Monograph No.1. Columbus, OH: Ohio State University.

Smith, A. and A. Venables. 1988. "Completing the Internal Market in the European Community-Some Industry Simulations." *European Economic Review* 32:1501-1525.

Sugden, R. 1983. "The Degree of Monopoly, International Trade, and Transnational Corporations." *International Journal of Industrial Organization* 1:165-187.

Sutton, J. 1991. *Sunk Costs and Market Structure*. Cambridge, MA: MIT Press.

Tirole, J. 1989. *The Theory of Industrial Organization*. Cambridge, MA: MIT Press.

Tyers, R. and K. Anderson. 1992. *Disarray in World Food Markets -- A Quantitative Assessment*. Cambridge: Cambridge University Press.

United Nations. 1990. "Statistical Papers, Commodity Trade Statistics, According to Standard International Trade Classification, Series D," Statistical Office, Department of Economic and Social Affairs.

Venables, A. and A. Smith. 1986. "Trade and Industrial Policy under Imperfect Competition." *Economic Policy* 3:622-672.

Verdoorn, P.J. 1960. "The Intra-Bloc Trade of Benelux," in Proceedings of a Conference Held by the International Economic Association, *Economic Consequences of Nations*, London.

2

International Trade, Market Structure, and Cyclical Fluctuations in U.S. Food Manufacturing

Martha K. Field and Emilio Pagoulatos

Introduction

This chapter examines the role of foreign trade, domestic market structure, and the business cycle on allocative efficiency (profit margins) using a panel of forty-three 4-digit U.S. food manufacturing industries from 1972 to 1987. This 16-year period spans several business cycles and enables examination of the impact of imports as imperfect substitutes for domestic goods on profit margins.

In light of increasing trade flows in the U.S. food industry (Reed and Marchant 1992), our model structure builds from several oligopoly models augmented to include the effect of foreign trade on domestic competition within an industry. Since Cowling and Waterson's (1976) seminal paper, theory has demonstrated the importance of demand elasticities as necessary elements in market performance models, especially models assuming imperfect competition. However, the theoretical inclusion of elasticities is more highly developed than empirical application because elasticities are generally unobservable and difficult to estimate. When considering an imperfectly competitive environment inclusive of international trade in imperfect substitutes, the problem of bias created by omitting elasticities intensifies, because trade elasticities augment those elasticities characterizing domestically produced goods (Huveneers 1981; Lyons 1981). Also, the influence of aggregate demand on market power cannot be overlooked; the position of an industry along its own and the national business cycle could impact the price-cost margin (PCM) independently of elasticities.[1]

The basic theoretical framework of oligopoly markets with international trade is developed in the next section, with a discussion of model specification and variable construction following. The statistical approach and empirical results are then presented, and the final section provides some concluding comments.

Theoretical Model

Although not originally in a trade setting, most theoretical modeling examining the impact of international trade variables on an industry's domestic market performance were developed from the Cowling and Waterson (1976) model, which demonstrated the importance of the price elasticity of demand (η_d) in market structure-performance models under the assumption of homogeneous goods. With international trade, the relation of imports to domestic production must be considered. In many competitive fringe models, imports are included in the domestic market as imperfect substitutes for domestically produced goods (DeRosa and Goldstein 1981; Huveneers 1981; Lyons 1981; Jacquemin 1982; Neumann, Böbel, and Haid 1985; Stalhammar 1991). Assume N domestic firms produce homogeneous goods that are differentiated from imports that are identical among themselves. There is no entry and the profit function for the ith firm is:

$$\Pi_i = P_d(D,M)D_i - AC(D_i)D_i \tag{1}$$

where Π_i is profit, D_i is output rate, AC equals average cost (Koh 1990), P_d is domestic price as the inverse market demand function of D and imports M, and $P_d=f(D,M)=f[\Sigma_{i=1}^n D_i, M]$. By maximizing (1) with respect to D_i and assuming competition is measured by conjectural variations (Lyons 1981) of the form:

$$\frac{\partial D}{\partial D_i} = 1 + \lambda_i$$

where

$$\lambda_i = \frac{\partial \Sigma_{j \neq i} D_j}{\partial D_i}.$$

The first-order condition for the ith firm is:

$$\frac{\partial \Pi_i}{\partial D_i} = P_d + D_i[f_D(1+\lambda_i) + f_M \frac{\partial M}{\partial D_i}] - c'(D_i) \tag{2}$$

Aggregating over N firms, and forming elasticities, generates the industry PCM:

$$PCM = \frac{H}{\eta_d}(1+\lambda) - \eta_{PM} e_{MD} \tag{3}$$

where H_d is the Herfindahl index, η_{PM} is the price flexibility of domestic goods with respect to imports,

$$\eta_{PM} = \frac{\partial P_d}{\partial M} \frac{M}{P_d}$$

and e_{MD} is the supply elasticity of imports with respect to domestic output,

$$e_{MD} = \sum_{i=1}^{n} \frac{D_i}{D} \frac{\partial M}{\partial D_i} \frac{D_i}{M}.$$

With product differentiation, imports and domestic goods contribute separate elasticities, which depend on the substitutability between domestic and foreign goods.

Using conjectural variations to reflect the response of imports to changes in domestic output has been criticized by Huveneers (1981) because it fails to consider the firm's size and the importer's profit function through which the import conjectural variation term must operate. Rather, importers respond to domestic price changes that are triggered by domestic output changes (Caves 1974). Huveneers (1981) replaced conjectural variations in Lyons' (1981) competitive fringe model with elasticities. In addition to supporting the inclusion of the import supply elasticity in explaining performance, the model rigorously demonstrated the importance of the substitutability between domestic and imported goods. Whereas Lyons (1981) initially assumed import quantity is set exogenously and considered only the price facing domestic firms, Huveneers (1981) assumed endogeneity of imports and, in addition to the domestic price function, included import price as a function of imports $[P_M = P_M(M)]$.

In a competitive fringe model, the set of dominant firms faces a residual demand, while the fringe determines quantity as a function of market price. Applying this to the import model, Huveneers (1981) defined import demand as a function of domestic price and import price $[M=M(P_d,P_M)]$. Incorporating these assumptions into equation (1) gives the objective function:

$$\prod_i = P_d[D,M(P_d,P_m)]D_i - AC(D_i)D_i. \qquad (4)$$

The first order condition for profit maximization is:

$$\frac{\partial \prod_i}{\partial D_i} = P_d + D_i[f_D + f_M(\frac{\partial M}{\partial P_d}\frac{\partial M}{\partial P_m}\frac{\partial P_m}{\partial P_d})f_D] - c'(D_i). \qquad (5)$$

By aggregating over N firms, forming elasticities, and by following Pugel (1980), Huveneers (1981) developed:

$$PCM = \frac{H_d}{\eta_d}[1 + \frac{\eta_{PM}\varepsilon_{MD}}{1 - \eta_m/\varepsilon_s}]. \qquad (6)$$

With D and M entering the market separately, elasticities from P_d and P_m are constructed separately as own price elasticity of domestic demand,

$$\eta_d = \frac{\partial D}{\partial P_d}\frac{P_d}{D}$$

and own price elasticity of imports,

$$\eta_m = \frac{\partial M}{\partial P_m}\frac{P_m}{M}$$

In equation (3) η_d included imports; in equation (6) η_d does not. Product differentiation contributes η_{PM} and, combined with the endogeneity of imports, yields the cross price elasticity of import demand,

$$\varepsilon_{MD} = \frac{\partial M}{\partial P_d}\frac{P_d}{M}.$$

The supply elasticity of imports,

$$\varepsilon_s = \frac{\partial S}{\partial P_m}\frac{P_m}{M}$$

is derived from the inverse of the P_m function $[M=S(P_m)]$.

With Huveneers' (1981) approach, own price and cross price elasticities of import demand enter the model and describe industry profitability. The cross elasticity combined with the elasticity of domestic price with respect to imports gives more weight to the degree of substitution between domestic and import goods. Greater substitutability between domestic goods and imports suggests a negative impact on profitability and a smaller but positive relationship between PCM and concentration. However, it is important to note that imports and profitability may be jointly determined, reflecting the feedback of trade on industry performance (Geroski and Jacquemin 1981; Jacquemin 1982; Chou 1986). Although endogeneity is not addressed in this chapter, simultaneity under imperfect competition with trade may be one possible extension of this study.

Model Specifications and Variable Construction

The theoretical model derived suggests two potential empirical specifications: one for the small country assumption and one for the large country assumption. In the former, the supply elasticity of imports, ε_s, is infinite, causing the ratio of own price elasticity of demand to the supply elasticity of imports, η_m/ε_s, to approach 0 in the limit. Hence, the PCM equation (6) becomes:

$$PCM = \frac{H_d}{\eta_d}(1+\eta_{pm}\varepsilon_{MD}). \tag{7}$$

The empirically estimable equation in double logarithmic functional form is:

$$\ln PCM = \beta_o + \beta_1 \ln H_d - \beta_2 \ln \eta_d + \beta_3 \ln(1+\eta_{pm}\varepsilon_{MD}) \tag{8}$$

For a large country, ε_s is no longer infinite, and the equation to be estimated is:

$$\ln PCM = \beta_o + \beta_1 \ln H_d - \beta_2 \ln \eta_d + \beta_3 \ln \frac{[1+\eta_{pm}\varepsilon_{MD}]}{1-\dfrac{\eta_m}{\varepsilon_s}} \tag{9}$$

In equations (8) and (9), the Herfindahl index (H_d) and the absolute value of the own price elasticity of domestic demand affect PCM positively and negatively, respectively. These predictions align with the predictions and results of previous empirical studies; Domowitz, Hubbard and Peterson (1986a,b) consider H_d and Pagoulatos and Sorensen (1981) examine η_d. Increased concentration could raise PCM via greater potential collusion from more interdependence among firms, thus generating greater industry market power. Greater price elasticity suggests more sensitivity to price changes, thereby generating a smaller PCM.

Interpretation of β_3 is more complicated. Although in general, the last terms represent the substitutability of differentiated imports for domestic goods within the same industry, the relationship among the elasticities dictates the two models' predictions. With the small country model, equation (8), the term $1+\eta_{PM}\varepsilon_{MD}$ has a positive effect on PCM. This prediction relies on the sign of the product of the cross elasticity of import quantity with respect to domestic price (ε_{MD}) and the domestic price flexibility with respect to imports (η_{pm}). This interdependence of domestic producers with imports affects PCM positively. This outcome may appear to challenge the established finding that imports act as a discipline on domestic market power. However, the positive effect may derive from potential collusion between domestic producers and importing firms (Urata 1979, 1984; Haubrich and Lambson 1986); alternatively, it may represent intrafirm imports (Sugden 1983). Also, the positive sign may suggest that industries that closely monitor import responses may closely scrutinize their own response, thus maintaining PCM.

The above arguments lead to the general form of the empirical model:

$$PCM = f(EL, CR, FT, BC) \tag{10}$$

where PCM is the price-cost margin, EL is the price elasticity of demand, CR is seller concentration, FT is foreign trade, and BC represents the influence of the business cycle.

Price-Cost Margins

Measuring market power via PCM can be quantified in several ways. By assuming PCM is observable, accounting data can be utilized and PCM calculated. Another approach assumes PCM is not observable but estimable from the production function. Lastly, the principles of duality can be employed to obtain demand conditions from the cost function (Hall 1988; Bresnahan 1989; Perloff 1992).

In this chapter, we assume that accounting rates of return approximate economic rates of return (Fisher and McGowan 1983). Also, because marginal costs are unobservable, we assume they equal average costs which are observable. Generally, PCM is measured as a ratio of value added minus payroll costs to total shipments (Urata 1979; Lyons 1981; DeGhellinck, Geroski, and Jacquemin 1988). Although the PCM measurement may be imperfect, it can sufficiently indicate changes and trends, and its consistency with other studies permits comparison of results. For this study, the price-cost margin is defined as:

$$PCM = \frac{\text{Value Added} - \text{Total Employment Payroll}}{\text{Value Added} + \text{Cost of Materials}}$$

Price Elasticity of Demand

While elasticities affect the ability of firms to change prices while retaining profitability, differences in elasticities reflect non-price competition, such as product differentiation, demand growth, collusive behavior, and other types of heterogeneity between industries (Schmalensee 1989a). Several valid assumptions can justify omission of elasticities.

First, if PCM is measured over time as a ratio between years, or by assuming constant elasticities, this causes the price elasticity of demand variable to drop mathematically from the estimated PCM equation (Cowling and Waterson 1976; Lyons 1981). This assumption is plausible within an industry over time but less so across industries. With trade, domestic demand can be assumed to approximate a perfectly elastic world demand (Stalhammar 1991). Then, the demand elasticity again drops from the equation. Alternatively, if one assumes a small country model, import elasticity could equal infinity because the country is a price taker on the world market (Jacquemin 1982). Yet, omission causes biased estimates, and a downward bias arises if the market demand elasticity is correlated with concentration (Clarke and Davies 1982).

Import market share has been used as a proxy for the unobservable domestic demand price elasticity and for the import supply elasticity under the assumption that these elasticities vary directly with import market share (DeRosa and Goldstein 1981). Low import share may reflect little stimulus for import entry, due to a lack of excess profits; with an elastic demand, any price change would not retain excess profits. Conversely, a high import share could be attributed to above normal profits and import entry, thereby representing a less elastic demand (Caves 1974). This argument could be accepted for a highly self-sufficient country with a relatively small trade component. It could characterize the U.S. food manufacturing sector prior to 1970, but it is inappropriate either for a heavily trade dependent country or for the globalized economies of today.

Estimation of elasticities is another approach. Mean price elasticities of demand can be estimated by weighting a price to consumption quantity ratio by the price coefficient of static consumer demand equations utilizing time series data (Pagoulatos and Sorensen 1986). From the following expression:

$$q = \beta_0 + \beta_1 P + \beta_2 Y + u \tag{11}$$

where q is an index of real per capita consumption, P is a price index, Y is an index of personal disposable income, and u is the error term, the elasticity is calculated as:

$$EL = \beta_1 \bar{\rho}$$

where $\bar{\rho}$ equals P/q, the mean of the price to consumption quantity ratio for the time period.

In this chapter, this regression methodology is adapted to capture changes in elasticities through the business cycle in addition to accounting for inter-industry differences. The business cycle could impact the price elasticity of demand through shifting the demand curve precipitated by demand growth or contraction. To maintain a constant elasticity with demand shifting, the demand curve must rotate in compensation. This may not occur through time. Hence, to capture some of the effects of the business cycle, yearly elasticities are calculated by redefining $\bar{\rho}$ as ρ_t, the ratio of price to consumption quantity for each time period (P_t/q_t):

$$EL_t = \beta_i \rho_t$$

Such yearly elasticities are calculated from a demand equation with a linear functional form that generates the own price coefficient, β_i, rather than the more commonly used linear logarithmic functional form which generates a constant own price elasticity. Demand for the ith industry appears as:

$$Q_i = \beta_{io} + \sum_{j=1}^{n} \beta_{ij} P_j + \beta_{iM} P_i^M + \beta_{iC} C + \beta_{it} T + u_i \qquad i,j=1,2,...,n \tag{12}$$

where Q_i is an index of deflated values of domestic shipments less the value of exports for industry i as a proxy for consumption quantity, P_j is a normalized industry shipment price index for goods in industry j, P_i^M is the normalized import price of goods in industry i, C is the real personal consumption expenditures per capita, T is year, and u_i is the error term. P_j includes the price of domestic substitute goods for the ith commodity, and j ranges from 1 to n where n equals the number of industries in the subsystem. Imported goods are assumed imperfect substitutes within industry i, and not substitutes for other imports. Hence, cross partials among imports are zero.

To ensure that the parameters of the estimated demand equation align with the foundation of demand theory, valid theoretical restrictions are applied which also generate greater statistical efficiency. More specifically, demand theory requires that conditions of additivity, homogeneity, and symmetry are satisfied.

The consumer demand equation (13) is estimated using annual data for 1972-87 for each food processing industry at the 4-digit standard industrial code (SIC) level.[2] All indices are based on 1982 dollars, and prices are normalized using the Consumer Price Index. All quantities are per capita relationships which have the advantage of being more stable than aggregate measures. For the sake of simplicity, we assumed consumers adjust consumption within the year, thus making equation (13) static. To maintain the focus on domestic production, we extracted exports from the value of shipments. Assuming that products in similar industries, such as meats and other protein-type products, are affected by the same random shocks captured in the error term, we grouped industries by related products to generate a seemingly unrelated regression model.

Although this is not a complex demand model, the objective is to obtain inter-industry differentials of price elasticities of demand. These estimates then become explanatory variables in a market performance regression model of the industry. Hence, the relationship of elasticity values to each other is more important than an accurate numerical value. The theoretical expectation is that the absolute value of EL will be inversely related to PCM.

Seller Concentration

The relationship of seller concentration to price-cost margins has been a cornerstone of many industrial organization studies. Two types of indices are commonly used for assessing the degree of market concentration: the four-firm concentration ratio (CR4) and the Herfindahl index (H_d).

In theoretical models, H_d for the domestic market should include imports. Insufficient data precludes such an accurate measure, and only domestic production is included empirically (Huveneers 1981; Neumann, Böbel, and Haid 1985; Stalhammar 1991).

Changes in the relationship of CR4 to PCM were explored for the 1970's and early 1980's (Domowitz, Hubbard, and Peterson 1986a, b; Salinger 1990). Salinger (1990) analyzed several measures of CR4, among them an adjusted CR4 multiplied by the complement of the imports (M) to sales (S) ratio, (1-M/S). In this case, as import share increases, the decline in concentration can be attributed to more sellers in the market. However, if imports equal sales then M/S approaches one and the adjusted concentration measure becomes zero in the limit. This latter case does not reflect domestic market concentration with a high degree of foreign competition.

The U.S. Census of Manufacturing compiles CR4 once every five years and makes it available for 1972, 1977, 1982, and 1987, while compilation of H_d only began in 1982. For non-census years, the CR4 utilized in this study was estimated by weighted averages of the preceding and succeeding censuses with the weights proportional to time differences (Salinger 1990). A second measure of concentration was also computed in order to capture the impact of foreign competition. The census CR4 was multiplied by the ratio of domestic values of shipments (VS) plus imports (M) to obtain the adjusted CRM = CR4 x VS/(VS + M). Thus, whenever imports are positive, CR4 is reduced. The theoretical expectation is that both CR4 and CRM are positively related to PCM.

International Trade

International trade has grown considerably for U.S. manufactured food products over the last two decades and has exceeded the increase in the value of domestic shipments. Export flows nearly tripled, and imports more than doubled during the 1972 and 1987 period. Thus, the overall export share for these products stood at about 3.7 percent in 1987, while import penetration reached 5.3 percent in 1987. These aggregate figures mask considerable variation among individual industries during the same period, but the overall trend has been one of steady increases in both export shares and import penetration for a large majority of industries in our sample.

To control for the degree of potential international competition in the empirical model would require estimates of the various import demand and supply elasticities implied by equation (3) of the theoretical model. The lack of available data prevents direct estimation of these elasticities. Therefore, we follow the approach suggested by Pugel (1980) and Jacquemin (1982) of including in the model an export intensity variable (X/VS) and an import penetration variable (M/DVS) where DVS = VS - X + M.

The effect of export intensity on PCM cannot be determined *a priori.* As Caves (1974, 1985) has pointed out, if domestic firms are unable to engage in price discrimination between the domestic and foreign markets, the existence of export markets may serve to constrain domestic industries to more competitive pricing behavior, implying a negative relationship between export share and PCM. This result also prevails if expansion of export sales weakens oligopolistic interdependence in the domestic market by flattening the demand curve facing the individual sellers. The share of exports in total sales, however, could be positively related to profitability if exporters, due to tariff protection, can engage in international price discrimination, if the industry enjoys international product differentiation, or if export sales, by increasing the sizes of plants and enterprises, lead to increased technical efficiency.

The relationship between import penetration and price-cost margins is also not entirely clear. In general, actual import competition can increase the number of competitors in the domestic market, in effect reducing domestic seller concentration and resulting in more competitively determined prices and lower profits for the domestic firms. The same result is expected from potential import competition, especially if foreign sellers are less impeded than potential domestic entrants by barriers to entry. Thus, imports can act as a source of competitive discipline in the domestic market and result in lower price-cost margins.

The possibility of a positive relationship between the price-cost margin and import share has been shown recently by several oligopoly models. Theoretical studies by Geroski and Jacquemin (1981), Urata (1984), and Haubrich and Lambson (1986) derive a positive relationship based on potential collusion between domestic producers and domestic firms. The same result is obtained by Sugden (1983) for the case of intra-firm imports by transnational corporations.

Business Cycle

The business cycle and changes in industry growth can impact trade and domestic performance. There are several theoretical underpinnings to demand growth considerations. Growth can facilitate entry of new firms, internal expansion of existing firms, or import shipments. Such growth in market size would not threaten PCM of incumbent firms. In the last case, the imports-as-market-discipline hypothesis may be supported because no firm would need to sacrifice to make room for the entrants. In the short run, capacity constraints may limit satisfying increased demand by stimulating price increases which raise PCM, or procyclical productivity gains can aid firms to capture scale economies that reduce costs, thus raising PCM (Domowitz, Hubbard, and Peterson 1986a, b).

Studies of overall U.S. manufacturing by Domowitz, Hubbard, and Peterson (1986a, b) and Salinger (1990) found that price-cost margins exhibited procyclical behavior during the 1970's, with the difference between concentrated and unconcentrated industries lessening through time. They found aggregate demand effects as one explanation for the intertemporal instability of PCM. Import competition contributed to the narrowing of the PCM gap, but procyclical behavior dominated the pattern. Also, it is possible that during a downturn, low market demand can trigger the collapse of collusive arrangements within oligopolies (Stigler 1964; Rotemberg and Saloner 1986). This would intensify competition and possibly limit PCM. Conversely, Schmalensee (1989b) and Bils (1987) found a counter cyclical relationship for U.S. manufacturing during the same period.

At the aggregate level, sectors of one industry face identical changes in macroeconomic growth variables such as gross domestic product, national capacity utilization, or unemployment levels. At the intraindustry level, sectors may be at different points along their growth paths in any given year. With some sectors shrinking while others are expanding, controlling for aggregate demand in a time series framework is vital. From a cross sectional view, growth can control for windfall profit differences (Stalhammar 1991).

To address the influence of demand changes and to capture the destabilizing effect of growth on performance in U.S. food manufacturing, this study uses three different variables. One is the growth rate of domestic industry output measured as the annual rate of change of industry value of shipments ($VS_{it} - VS_{it-1}/VS_{it-1}$). Second, to capture the national business cycle, we alternatively use the annual aggregate unemployment rate (U_t) and aggregate capacity utilization in manufacturing (CU_t).

Statistical Approach and Results

The data set used here consists of N=43 cross-sectional units observed at each of T time periods, where T equals 16 annual observations from 1972 through 1987.[3] Given the dimensions of this panel, cross-sectional and time-series effects are an integral consideration of the econometric specification. This panel is a self-contained population rather than a random sample drawn from a population, and conditional inferences are the focus of this investigation. Also, nonorthogonality between omitted variables and some of the explanatory variables may exist. These factors point toward a fixed-effects model, even though N is relatively larger than T. However, for this preliminary portion of the study, we employ a generalized least squares model with pooled data and will address the fixed effects model at a later date.

The pooled data set was tested for violations of econometric assumptions. These tests revealed mild signs of heteroscedasticity and autocorrelation and no problem with multicollinearity. A Goldfeld-Quandt test rejected the null hypothesis of homoscedasticity. First order autocorrelation failed to be rejected in 11 of the 43 industries using a Durbin-Watson test. A low variance inflation factor, 1.1, combined with the determinant of the $X'X$ matrix equal to 0.75, indicated that multicollinearity was not a problem. Considering these findings, we used a cross-sectionally heteroscedastic and time-wise autoregressive model for the pooled data set.

The generalized least squares results for four alternative specifications of the price-cost margin model are presented in Table 2.1. The values in parentheses underneath each coefficient estimated are standard errors. The results are quite robust in that all of the estimated regression coefficients are statistically significant.

The significance of market price elasticity of demand in inversely affecting price-cost margins confirms the importance of this variable. Also conforming to theoretical expectation was the positive and significant effect of seller concentration on margins. This result was unaltered when the four-firm concentration was adjusted by imports.

The results for the import share variable suggest that its impact on margins is not only direct, but also through its effect on adjusting concentration. More specifically, the imports to domestic value of shipments ratio shows a positive and significant effect on margins in our sample of U.S. food manufacturing industries. This result is consistent with the view emphasizing intra-firm imports and collusion between domestic and foreign producers rather than the imports-as-discipline view. Indirect evidence of this can be found in studies that detail the relatively large presence of foreign direct investment inflows and outflows and intra-firm trade in U.S. food processing (Handy and Henderson 1992).

A further interesting result is that the coefficient of the export share is negative and statistically significant at the 1 percent level. This finding confirms the hypothesis that industries characterized by a high rate of exports leave scope for a larger number of domestic competitors and reduced margins, especially when the domestic market cannot be segregated from the world market.

TABLE 2.1   Regressions on Price-Cost Margins for 43 Food Manufacturing Industries[*]

Exogenous Variables	(1)	(2)	(3)	(4)
Constant	-3.572[**]	-3.426[**]	-3.311[**]	-3.171[**]
	(.156)	(.166)	(.171)	(.182)
log EL_{it}	-.076[**]	-.076[**]	-.077[**]	-.078[**]
	(.019)	(.020)	(.020)	(.020)
log $CR4_{it}$	.480[**]		.480[**]	
	(.036)		(.036)	
log CRM_{it}		.456[**]		.455[**]
		(.039)		(.038)
UN_t	.0098[**]	.0094[**]		
	(.0033)	(.0034)		
CU_t			-.0024[**]	-.0023[**]
			(.00088)	(.00088)
VS_{it}-VS_{it-1}/VS_{it-1}	.062[***]	.062[***]	.058[***]	.058[***]
	(.028)	(.028)	(.028)	(.028)
log $(M/DVS)_{it}$	.020[***]	.029[**]	.020[***]	.029[**]
	(.0087)	(.0093)	(.0087)	(.0093)
log $(X/VS)_{it}$	-.057[**]	-.057[**]	-.057[**]	-.057[**]
	(.0082)	(.0084)	(.0082)	(.0084)

Note: [*] Dependent Variable is log PCM_{it} - i and t denote the industry and time period, respectively; standard errors are in parentheses; pooled time series-cross section over the 1972-1987 period with 688 total observations.
[**] Indicates significant at the 1 percent level.
[***] Indicates significant at the 5 percent level.

Turning next to the business cycle variables, we observe that the aggregate unemployment rate is positively related with price-cost margins, while the aggregate capacity utilization in manufacturing has a negative effect on margins. Both variables are statistically significant at the 1 percent level and suggest a counter cyclical response of margins in U.S. food manufacturing industries. This result indicates that the profit advantage of firms in the food industries is greater in recessions than in boom periods. This outcome is consistent with both the work of Bils (1987), who finds that marginal cost is procyclical, and the work of Mills and Schumann (1985), who found that large firms lose fewer sales during recessions and expand output less in booms than their smaller competitors. While these explanations are useful, they are not conclusive and more work needs to be done in further understanding the counter cyclical nature of margins in U.S. food industries.

The final result considered in Table 2.1 is the positive and significant coefficient for the annual growth in industry value of shipments variable. This fairly traditional outcome suggests that margins respond procyclically to the growth in nominal output in the industry.

Conclusions

This chapter provides new evidence on the relationship between price-cost margins, foreign trade, domestic market structure, and the business cycle for a panel of forty-three U.S. food manufacturing industries from 1972 and 1987. The basic model specification and variables are derived from static oligopoly theory.

Our key findings support the negative influence of price elasticity of domestic demand on margins and the important role of seller concentration in exerting a positive effect on price-cost margins in the industry. The opening to foreign trade also yields some interesting results. The positive relationship between import share and margins and the negative effect of exports suggest that it is expanding exports that results in competitive pressure from abroad rather than increasing import penetration. Finally, regarding the business cycle, price-cost margins exhibit counter cyclical behavior at the national level while behaving procyclically at the industry level.

Notes

1. The price-cost margin (PCM) measures the difference between selling price and marginal cost, approximating the Lerner index of monopoly. Hence, it can be assumed to measure market power. A limitation of this assumption is that PCM can reflect cost efficiency differences between firms (Demsetz 1974). Although debate is not closed on this issue, both explanations reflect reduced competition.

2. Data used for the estimation of equation (13) came from a variety of sources. The main consideration was to obtain a consistent time series for each industry. Primary reliance was on the *1989 U.S. Industrial Outlook Data File* supplied by the International Trade Administration, U.S. Department of Commerce. These data were supplemented with information obtained from U.S. Department of Agriculture publications: *Food Consumption, Prices and Expenditures*; *Agricultural Statistics*, and *Foreign Agricultural Trade of the U.S.* (several issues).

3. Data were obtained primarily from the *1989 U.S. Industrial Outlook Data File* supplied by the International Trade Administration, U. S. Department of Commerce. The 1987 SIC numbers of the industries included in this study are: 2011, 2013, 2016, 2017, 2021. 2022, 2023, 2024, 2026, 2032, 2033, 2034, 2035, 2037, 2038, 2041, 2043, 2044, 2045, 2046, 2048, 2051, 2052, 2061, 2062, 2063, 2065, 2066, 2067, 2074, 2075, 2077, 2079, 2082, 2083, 2084, 2085, 2086, 2087, 2092, 2095, 2097, and 2098. Additional data sources were the *Economic Report of the President 1989* and the *U. S. Department of Commerce, Bureau of the Census, 1987 Census of Manufactures: Concentration Ratios in Manufacturing*, Washington, D. C. 1991.

References

Bils, M. 1987. "The Cyclical Behavior of Marginal Cost and Price." *American Economic Review* 77:838-855.

Bresnahan, T.F. 1989. "Empirical Studies of Industries with Market Power," in R. Schmalensee and R. D. Willig, eds., *Handbook of Industrial Organization*. Amsterdam: North Holland.

Caves, R.E. 1974. "International Trade, International Investment and Imperfect Markets." *Selected Papers in International Economics*. Princeton, NJ: Princeton University.

_____. 1985. "International Trade and Industrial Organization: Problems, Solved and Unsolved." *European Economic Review* 28:377-395.

Chou, T. 1986. "Concentration, Profitability and Trade in a Simultaneous Equation Analysis: The Case of Taiwan." *The Journal of Industrial Economics* 34:429-443.

Clarke, R. and S.W. Davies. 1982. "Market Structure and Price-Cost Margins." *Economica* 49:277-287.

Cowling, K. and M. Waterson. 1976. "Price-Cost Margins and Market Structure." *Economica* 43:275-286.

De Ghellinck, E., P.A. Geroski and A. Jacquemin. 1988. "Inter-Industry Variations in the Effect of Trade on Industry Performance." *Journal of Industrial Economics* 37:1-19.

Demsetz, H. 1973. "Industry Structure, Market Rivalry and Public Policy." *Journal of Law and Economics* 16:1-10.

DeRosa, D. and M. Goldstein. 1981. "Import Discipline in the U.S. Manufacturing Sector." *International Monetary Fund Staff Papers* 28:600-634.

Domowitz, I., R.G. Hubbard, and B.C. Peterson. 1986a. "Business Cycles and the Relationship Between Concentration and Price-Cost Margins." *Rand Journal of Economics* 17:1-17.

_____. 1986b. "The Intertemporal Stability of the Concentration Margins Relationship." *Journal of Industrial Economics* 35:13-34.

Fisher, F. and J.J. McGowan. 1983. "On the Misuse of Accounting Rates of Return to Infer Monopoly Profits." *American Economic Review* 73:82-97.

Geroski, P.A. and A. Jacquemin. 1981. "Imports as a Competitive Discipline." *Recherches Economiques de Louvain* 47:197-208.

Hall, R.E. 1988. "The Relation Between Price and Marginal Cost in U. S. Industry." *Journal of Political Economy* 96:921-947.

Handy, C.R. and D.R. Henderson. 1994. "Foreign Investment in Food Manufacturing." NC-194 Occasional Paper OP-41. Columbus, OH: Ohio State University.

Haubrich, J. and V.E. Lambson. 1986. "Dynamic Collusion in an Open Economy." *Economics Letters* 20:75-78.

Huveneers, C. 1981. "Price Formation and the Scope for Oligopolistic Conduct in a Small Open Economy." *Recherches Economiques de Louvain* 47:209-242.

Jacquemin, A. 1982. "Imperfect Market Structure and International Trade - Some Recent Research." *Kyklos* 35:75-93.

Koh, Y. 1990. "Market Structure, Profitability and Trade Performance: The Case of the Korean Manufacturing Industry." Unpublished Doctoral Dissertation, Fordham University.

Lyons, B. 1981. "Price-Cost Margins, Market Structure and International Trade, "in D. Currie, D. Peel, W. Peters, eds., *Microeconomic Analysis: Essays in Microeconomic and Economic Development*. London: Croom Helm.

Mills, D. E. and L. Schumann. 1985. "Industry Structure with Fluctuating Demand." *American Economic Review* 75:758-767.

Neumann, M., I. Böbel, and A. Haid. 1985. "Domestic Concentration, Foreign Trade and Economic Performance." *International Journal of Industrial Organization* 3:1-19.

Pagoulatos, E. and R. Sorensen. 1981. "A Simultaneous Equation Analysis of Advertising, Concentration and Profitability." *Southern Economic Journal* 47:728-741.

_____. 1986. "What Determines the Elasticity of Industry Demand?" *International Journal of Industrial Organization* 4:237-250.

Perloff, J.M. 1992. "Econometric Analysis of Imperfect Competition and Implications for Trade Research," in I.M. Sheldon and D.R. Henderson, eds., *Industrial Organization and International Trade: Methodological Foundations for International Food and Agricultural Market Research*, NC-194 Research Monograph No. 1. Columbus, OH: Ohio State University.

Pugel, T.A. 1980. "Foreign Trade and U.S. Market Performance." *Journal of Industrial Economics* 29:119-129.

Reed, M.R. and M.A. Marchant. 1992. "The Global Competitiveness of the U. S. Food-Processing Industry." *Northeastern Journal of Agricultural and Resource Economics* 21:61-70.

Rotemberg, J.J. and G. Saloner. 1986. "A Supergame-Theoretic Model of Price Wars during Booms." *American Economic Review* 76:390-407.

Salinger, M. 1990. "The Concentration-Margins Relationship Reconsidered." *Brookings Papers on Economic Activity: Microeconomics* 30:287-335.

Schmalensee, R. 1989a. "Inter-Industry Studies of Structure and Performance," in R. Schmalensee and R.D. Willig, eds., *Handbook of Industrial Organization*. Amsterdam: North-Holland.

_____. 1989b. "Intra-Industry Profitability Differences in U. S. Manufacturing 1953-1983." *Journal of Industrial Economics* 37:337-357.

Stalhammar, N. 1991. "Domestic Market Power and Foreign Trade." *International Journal of Industrial Organization* 9:407-424.

Stigler, G.W. 1964. "A Theory of Oligopoly." *Journal of Political Economy* 72:44-61.

Sugden, R. 1983. "The Degree of Monopoly, International Trade, and Transnational Corporations." *International Journal of Industrial Organization* 1:165-187.

Urata, S. 1979. "Price-Cost Margins and Foreign Trade in U.S. Textile and Apparel Industries: An Analysis of Pooled Cross-Section and Time-Series Data." *Economics Letters* 4:279-282.

_____. 1984. "Price-Cost Margins and Imports in an Oligopolistic Market." *Economics Letters* 15:139-144.

3

Price Discrimination by U.S. High-Value Food Product Exporters

Paul M. Patterson, Alejandro Reca, and Philip C. Abbott

Introduction

A growing number of theoretical models proposed under the New Trade Theory seek to explain trade patterns and firm behavior under varying forms of imperfect competition (Helpman and Krugman 1989). These theoretical models have been motivated by a number of stylized facts on the nature of competition in international markets. For example, it has been observed that trade often occurs between a few large buyers and sellers. This is notably the case for agricultural products, where the number of government marketing boards with monopoly power over international trade is extensive. Even in nations where there are no state trading agencies, international trade is often handled by a few large firms (Morgan 1979; Patterson and Abbott 1992). Beyond anecdotal and institutional evidence, there is relatively little empirical evidence on aspects of imperfect competition in international markets, much less international food markets. Furthermore, the new theoretical models do not lend themselves easily to empirical testing. Thus, there has been a call for additional empirical evidence on imperfect competition in international markets using alternative models and tests.

This call for additional empirical evidence partially motivated much of the recent interest in examining export pricing behavior. Krugman (1987) suggested that exporters may practice a form of third degree price discrimination (Pigou 1920) across export markets in response to exchange rate movements. He called this strategy "pricing to market" (PTM). He argued that PTM may occur if an exporter either holds his domestic currency export price constant or raises (lowers) it for an importer who realizes a currency appreciation (depreciation). This has the effect of either allowing the importer's domestic currency price to fall (by proportionally less than the exchange rate change) or maintaining a stable import price.

Knetter's (1989) proposed empirical approach has proven to be a particularly popular method for testing the PTM hypothesis. Knetter (1989) analyzed the pricing behavior of U.S. and German exporters for a number of products, including a few food products -- onions, bourbon, orange juice, and breakfast cereals. Knetter (1989) found little evidence of PTM by U.S. exporters. He suggested that German exporters were more inclined to use such a strategy. In their application of Knetter's (1989) model to U.S. cotton, wheat, corn, soybeans, and soybean meal exports, Pick and Park (1991) also found little evidence of PTM by U.S. agricultural product exporters, except in the wheat export sector.

Other empirical efforts to link imperfect competition and international trade can be found in some of the early structure-conduct-performance (SCP) studies (see for example Pagoulatos and Sorenson 1976). These studies analyzed how import competition and export trade affected the profitability[1] of domestic industries. Lyons (1981) showed that an industry's profitability should be measured as a weighted average of its profitability in the domestic and export markets and that this weighted average measure of profitability is a function of seller concentration ratios in the domestic and export markets.

Due to data limitations, Lyons' (1981) model had not been tested until the recent study by Patterson and Abbott (1992). They found that export seller concentration and market share do significantly affect destination-specific wheat and corn export prices, hence the pattern of price discrimination. It remains to be seen whether these structural factors influence the pattern of prices observed in export markets for other food products.

The purpose of this chapter is to test for discriminatory pricing behavior in two U.S. high value food product export markets -- chicken and beef. The model proposed by Knetter (1989) and an adaptation of Lyons' (1981) model will be estimated. The estimation results provide evidence on the extent of price discrimination for these products and on factors determining the observed pattern of price discrimination. This analysis also evaluates the effect of product aggregation on tests of export price discrimination.

U.S. chicken and beef exports have experienced significant growth in recent years. Also, the level of concentration in domestic and export sales is relatively high. Seller concentration in the U.S. beef market has been a concern throughout the 1980's (Schroeter 1988). Recent mergers in the chicken processing sector have pushed the domestic four firm concentration ratio (CR4) past the 40 percent level (Rogers 1992). U.S. export seller concentration in these export markets is quite high, with the

CR4 reaching 100 percent to some destinations (Patterson, Reca, and Abbott 1993). This suggests that there may be potential for noncompetitive behavior in export markets.

Models of Price Discrimination

The assumptions and derivations of the theoretical and empirical models used to test for price discrimination in export markets are briefly reviewed below. The PTM model is explained first. Then an export oligopoly model following the SCP tradition is presented.

Knetter's Pricing to Market Model

Knetter's (1989) model is one of several that has attempted to test Krugman's (1987) PTM hypothesis. His is derived from a static monopoly profit maximization problem. Others have considered dynamic models of PTM (or exchange rate pass-through[2]), where either supply side factors (marketing distribution costs; Baldwin and Foster 1986) or demand side factors (consumer switching costs; Froot and Klemperer 1989) affect the rate and extent of adjustment in export prices in response to bilateral exchange rate changes. If an exporter is required to make costly investments in foreign distribution systems that are fixed in capacity, it may attempt to maintain a stable foreign currency price (price to market) in the face of exchange rate moves that are viewed as temporary. Alternatively, if current market share is a function of past market share (as empirical evidence indicates), suggesting that consumers face switching costs, then exporters may seek to maintain a stable foreign currency price during an expected temporary appreciation of their currency. Alternatively, they may allow the foreign currency price to fall during a currency depreciation to expand market share.

While the dynamic models may provide a richer characterization of firm strategies, Knetter's (1989) static model has proved to be popular in empirical studies because of its simplicity in formulation and empirical testing. It requires data that are readily available -- export prices (or unit values) and exchange rates. Given its popularity and simplicity, Knetter's (1989) model will be used here as one test for export price discrimination.

From the profit maximization problem for a monopoly exporter facing some residual export demand, Knetter (1989) derives the usual monopoly markup relationship:

$$P_{jt} = c_t\left(\frac{\varepsilon_{jt}}{\varepsilon_{jt} - 1}\right) \quad j=1,...,N \quad t=1,...,T. \tag{1}$$

P_{jt} is the monopolist's export price to destination j; c_t is the marginal cost of production in period t, which is assumed constant across all destinations; and ε_{jt} is the elasticity of the residual demand in market j facing the monopolist in terms of the importer's local currency price. Thus, the export price is a markup over marginal cost. This markup equation is consistent with either perfect or imperfect competition. If the market is competitive, then ε_{jt} is infinite and price equals marginal costs. The less elastic the residual demand in market j, the greater the markup. Changes in the exchange rate can bring about changes in the residual demand elasticity.

Knetter (1989) then proposes an empirical model capable of testing for three forms of behavior in the export market -- competitive behavior and two forms of imperfect competition. A fixed-effects regression model applied to a pooled, cross-sectional data set is used:

$$\ln p_{jt} = \theta_t + \lambda_j + \beta_j \ln s_{jt} + v_{jt} \quad j=1,...,N \quad t=1,...,T. \tag{2}$$

θ_t is a time period dummy variable; λ_j is a country dummy variable; s_{jt} is the bilateral exchange rate expressed in foreign currency units per exporter's currency (e.g. FCU/$); and, v_{jt} is a random disturbance term.

Three forms of market behavior may be inferred from estimated parameters. One, the null hypothesis of competition is tested by the case where all λ_j and all β_j are zero. In this case, prices equal marginal cost and export prices are the same for all destinations. Changes in marginal costs over time are captured by the time dummy variable, θ_t. Thus, there is no residual variation in the data that could be accounted for by either the country effects or exchange rates. Two, price discrimination across markets is allowed, but a constant elasticity residual demand schedule is assumed for each market. Here, λ_j is allowed to differ across markets, but the β_j's are zero. Three, price discrimination is allowed across markets, but the elasticity of residual demand is no longer assumed to be constant. In this case, a change in the exchange rate causes a change in the local currency price. With the elasticity of residual demand assumed to be nonconstant, the change in the importer's local currency price causes a change in the elasticity, and thus, a change in the markup. In cases where demand is less convex than a constant elasticity of demand schedule, β_j is expected to be negative, indicating PTM. However, if the demand schedule is more convex than the constant elasticity demand schedule, the profit maximization solution indicates that a positive β_j should be expected.

While the theoretical model allows for either a positive or negative exchange rate coefficient estimate, a negative coefficient is consistent with the strategy proposed by Krugman (1987). In this case, a relatively stable price is maintained in the foreign markets. In cases where β_j is positive, the exporting firm is magnifying the exchange rate effect by raising its export price for an importer who has already experienced a currency depreciation.

It should be noted that proper tests for the two forms of imperfect competition require data for individual products. If the data were composed of a mix of products (an aggregation) with varying unit values, and if import demand parameters varied across the products, then one could wrongly conclude that prices differed across destinations due to some form of price discrimination. This difficulty plagues empirical tests using readily available data. Issues on product aggregation in the data will be explored later in this chapter.

Knetter's (1989) empirical model is an ideal test for a country where export trade is monopolized, such as one employing a state trading agency, which is common in world agricultural trade. In that case, the available data are consistent with the theoretical model. However, it is unclear how Knetter's (1989) model extends to an analysis of export pricing for a sector composed of noncooperative oligopolists, unless they collude. Knetter (1989) acknowledges this problem and suggests that the model is still applicable in cases where either the products supplied by different firms are sufficiently differentiated or the foreign destination markets are segmented among firms, so that strategic interactions among firms do not arise. This argument is a bit perplexing, since PTM may indeed be a strategic reaction by firms.

Knetter's (1989) model may provide some evidence on the existence of discriminatory pricing, but it does not provide evidence on what factors explain the observed pattern of discriminatory pricing. This suggests that the traditional oligopoly models that link market structure variables to pricing behavior may be useful for analysis of export pricing.

Export Oligopoly Model

Lyons (1981) argued that industry profitability is a weighted average of the domestic and export price-cost margin. This weighted average price-cost margin is a function of concentration, measured by the Herfindahl-Hirschman index (HHI), in both the domestic and export market. Lyons' (1981) model will be adapted here to consider only the link between export pricing and export market structure. This gives rise to a price markup relationship for oligopolists in a particular country (e.g. the U.S.) supplying several export markets (see Patterson, Reca and Abbott 1993 for derivation):

$$\frac{P_{jt} - \overline{c}_t}{P_{jt}} = \frac{HHI_{jt}}{e_{jt}}[1 + B_j + \Phi_j]\frac{Q_{jt}}{(Q_{jt} + Q_{jt}^*)} \quad j=1,...,N, \ t=1,...,T \tag{3}$$

The term on the left hand side is the familiar Lerner index, where $\overline{c}_t$ is an industry average marginal cost of supplying all export markets. HHI_{jt} is the Herfindahl-Hirschman index of exporter concentration in market j; e_{jt} is the import demand elasticity in market j (the elasticity of the excess demand schedule); B_j and Φ_j are weighted average conjectures, where the weights correspond to firm market shares. In the current analysis, B_j is the average conjecture each U.S. firm holds on the quantity response of its U.S. rivals to a change in its output. Similarly, Φ_j is the average U.S. firm conjecture on the output supply response of foreign firms. The last term on the right hand side is the share of market j supplied by U.S. firms. Q_{jt} is the quantity supplied by all U.S. firms; Q_{jt}^* is the quantity supplied by all other foreign firms. This expression indicates that the markup over marginal cost increases with exporter concentration, increases with the exporters' share in market j ($Q_{jt}/[Q_{jt}+Q_{jt}^*]$), and decreases with the import demand elasticity.

The model specified in equation (3) is similar to one derived by Cowling and Waterson (1976). The industrial organization studies that followed the Cowling and Waterson (1976) specification typically extended their model by including other control variables in an additive form. Other control variables considered here include: (1) the total volume of U.S. exports of the product; (2) the real national income of import country j; and (3) a variable controlling for U.S. export policies. The export volume variable tests whether markups are higher during periods of export growth or during periods of decline. Real national income is included to account for the possibility that exporters may price discriminate on the basis of income (second degree price discrimination; Pigou 1920). This possibility was recently suggested by Knetter (1992).

The extended form of equation (3) follows in the tradition of the SCP models used in the analysis of domestic markets. The weaknesses of this class of models are well known (Bresnahan 1989). However, these models have seen limited use in the analysis of export trade, and have served to establish some important empirical regularities on domestic markets. Thus, in spite of its weaknesses, this type of analysis may prove to be a useful first step.

There are some similarities in the two models presented. Both models assume that the appropriate geographic markets are defined by foreign country destinations. Therefore, pricing behavior is analyzed along broadly defined trade routes. Also, the derivation of each theoretical model is similar. However, they differ in terms of the import demand schedule which is used and the nature of competition which is assumed. Knetter's (1989) model presumes that a profit maximizing monopolist (or cartel) faces some residual demand for its product. Thus, strategic interactions between domestic exporting firms and between the domestic and foreign firms are presumed not to occur (or are implicit in the residual demand function). In contrast, Lyons' (1981) model assumes that domestic exporting firms hold conjectures on the response of domestic and foreign firms to changes in their export quantities. The demand schedule corresponds to the importing country excess (net import) demand schedule.

Empirical Results

The estimation results from Knetter's (1989) PTM model and the export oligopoly model are presented in this section. First, the PTM model and export oligopoly model results for U.S. chicken exports are presented. Then results from applying the PTM model to U.S. beef product exports are presented. The models for beef use highly disaggregated data, which are only available since the United States' adoption of the Harmonized Trade Codes in 1989. Due to product quality differences, it has been argued that more refined product data are needed in analyses of beef trade (Hayes *et al.* 1991). The estimated beef models allow for an assessment of the affects of commodity aggregation on tests on price discrimination. The export oligopoly model is not estimated for beef since it would require that highly aggregated beef trade data be used.

Knetter's Pricing to Market Model and U.S. Chicken Exports

The exchange rate data needed to estimate Knetter's (1989) model were obtained from the International Monetary Fund's *International Financial Statistics* and were expressed in terms of the importer's currency per U.S. dollar (FCU/$U.S.). The nominal exchange rate values were converted to real values by deflating each country series by the price level (Producer Price Index or Consumer Price Index) in each destination country. Further, these real exchange rates were normalized on a given period. Country destinations which showed no variation in the nominal exchange rate during the sample period were dropped from the sample. Export unit values were used as proxies for the export price variable and were calculated using the U.S. Department of Commerce export value and quantity data available by destination (trade route).

The ready availability of monthly exchange rate data and export price data requires the researcher to make a decision on the appropriate level of temporal aggregation. Previous studies using the Knetter (1989) model have typically used quarterly data. A bivariate time series model for exchange rates and export prices and Granger causality tests were used here to assess the dynamic relationship between these variables (Bessler and Brandt 1982). It was found that exchange rates over a lag period of about two months influence chicken export prices. Thus, quarterly data seems appropriate.

The Knetter (1989) model was estimated by ordinary least squares using pooled cross-sectional, time series data sets for the period 1978 through 1991.[3] Typically one time dummy and one country dummy are excluded from the model to avoid singularity problems. The estimated coefficients for the country dummies become the difference between the country effect for the respective country and the country effect for the excluded country during the excluded year. However, these coefficient estimates offer no real evidence on the nature of competition and are sensitive to which country is excluded from the model. Suits (1984) suggested that the interpretation of dummy variable models can be enhanced by imposing the restriction that all dummy coefficients sum to zero (or some other constant). In this case, the country dummies measure the difference between the country effect for the respective country and the "national average" export price.[4] This parameter estimate indicates destinations where discriminatory prices were offered. Thus, useful information on the nature of competition is provided. With no time or country dummies excluded, T time, N country, and N exchange rate coefficients are estimated using samples containing TxN observations.

Here, chicken exports are defined to be composed of shipments of chicken parts and whole chickens.[5] The majority of U.S. chicken exports are in the form of chicken parts. While the chicken aggregate used here contains a great deal of product diversity, it is frequently used in reporting trade and production statistics and it has been used in previous trade studies (Alston and Scobie 1987). Inferences on the behavior of export agents are still appropriate, since the same group of firms is engaged in the trade of all these products.

Table 3.1 presents the results from the Knetter (1989) PTM model for chicken. The null hypothesis that chicken export prices to all destinations are equal ($H_0:\lambda_i=\lambda_j$) is rejected at the five percent level of significance (F-test). Also, the estimated country effect dummies point to a number of markets where discriminatory prices were found (Canada, Mexico, Netherlands Antilles, Colombia, Netherlands, Hong Kong, and Japan). These results should be evaluated with some caution given the heterogeneity of this commodity aggregate. In short, "parts are not parts." The variability in export prices across destinations may be due to differences in product form (Thornton 1993).

The hypothesis that all exchange rate coefficients are equal to zero ($H_0 : \beta = 0$) is rejected at the five percent level. Significant negative exchange rate coefficients are found for Canada and the Netherlands, thus supporting the PTM hypothesis in those markets.

TABLE 3.1 Knetter Pricing to Market Model for U.S. Chicken Exports

Country	Country Effect (λ)	Exchange Rate (β)
Canada	0.229** (10.967)	-0.876** (-6.257)
Mexico	-0.103** (-4.208)	-0.137 (-1.130)
Netherlands Antilles	0.147** (6.679)	-0.159 (-1.245)
Colombia	-0.072** (-2.882)	0.212 (1.219)
Netherlands	-0.203** (-4.773)	-0.873** (-7.261)
Singapore	-0.026 (-0.741)	-0.229 (-1.211)
Hong Kong	-0.121** (-5.402)	-0.125 (-0.886)
Japan	0.140** (4.790)	0.088 (0.796)
F-Test ($H_0 : \lambda_i = \lambda_j$)	35.679**	
F-Test ($H_0 : \beta = 0$)		11.373**

Note: The values in parentheses are t-values. One and two asterisks (* and **) denote significance at the ten and five percent levels, respectively.

The Export Oligopoly Model and U.S. Chicken Exports

The extended form of equation (3) specified the price-cost margin or Lerner index as a function of the ratio of the export Herfindahl-Hirschman index and the import demand elasticity, the U.S. market share in the destination market, and other control variables. A proxy for the Lerner index was used in this study. The price component was measured by export unit values, as in the previous section. The U.S. domestic wholesale price for chicken served as a proxy for marginal costs. Prices have been used as a proxy for marginal costs in other studies (see Connor and Peterson 1992). Since chicken meat includes shipments of both chicken parts and whole broilers, the domestic price was calculated as a weighted average of the domestic price for these product forms with the weights corresponding to the export volumes of these products to specific destinations.[6] The U.S. wholesale price of chicken leg quarters and whole broilers (Perez, Weimar, and Cromer 1991) was used as measures of the domestic price for chicken parts and broilers, respectively.

U.S. seller concentration in foreign country markets was measured by the Herfindahl-Hirschman index (HHI). This variable was calculated for each destination and year using the *Journal of Commerce* Port Export/Import Reporting Service data (Patterson and Abbott 1991). The HHI was weighted by the elasticity of import demand (e_j) for each destination. These elasticities are reported in Patterson, Reca, and Abbott (1993). The import demand elasticities were assumed to be constant over time. The U.S. market share variable (USMS) was measured using official USDA data (Gundsman and Webb 1991). National income (Y) and price level (P) data were obtained from *International Financial Statistics*. Information on the Export Enhancement Program (EEP) was obtained from reports provided by the USDA's Foreign Agricultural Service.

Data on the conjectures held by U.S. firms were not available and no attempt made to estimate them. Thus, Cournot behavior is implicitly assumed. Also, the model was estimated with the market share variable entering in an additive form, rather than as an interaction term with the concentration measure. The parameter estimates and t-values (in parentheses) are presented below:

$$\mathcal{L} = -0.449 + 3.906\,HHI/e - 0.388USMS + 9.259E-5Q + 0.001Y/P - 4.999EEP$$
$$(-0.037)\quad (2.035)\qquad (-4.535)\qquad (2.048)\qquad (2.224)\qquad (-0.255)$$

$$\overline{R}^2 = 0.15 \quad F\text{-}Value = 6.996 \quad No.Obs = 171 \tag{4}$$

It can be seen that the export markup is significantly and positively related to concentration at the 5 percent level, as expected. However, contrary to prior expectations, the markup is negatively correlated with U.S. market share in the import markets. There are several possible explanations for this result. One, U.S. market share has expanded in some markets through increased exportation of lower valued chicken parts. Thus, the export unit value has fallen by more than is reflected by the proxy measure for marginal costs. Two, U.S. exporters lowered export prices in order to expand market share. Three, U.S. exporters have lost market share in some markets, but have maintained their export price markup. Each of these corresponds to a strategic behavior by firms different from that presumed in the model.

It can be seen that export volume does have a slight, positive influence on the price markup. Therefore, higher markups may be earned in growth export markets. It is also seen that a higher real national income may increase the markup. This variable accounts for variations in national income not captured by the import demand elasticity, which was assumed to be constant over time. The coefficient for the dummy variable, introduced to control for markets and years where the EEP was implemented, is negative, but not significantly different from zero. It is noted that the EEP was used primarily for whole broilers and only in a few markets. Therefore, the policy had a minimal impact on the observed unit value for chicken meat exports.

The results from the alternative models suggest that: (1) U.S. chicken export prices differ across destinations; (2) Movements in exchange rates appear to influence the export price to some destinations; (3) Export seller concentration and importer income are positively correlated with the export price markup, hence the pattern of price discrimination.

In an effort to reconcile the results obtained from the two alternative models, another model was estimated. The Knetter (1989) PTM model was extended by including the Herfindahl-Hirschman index as an explanatory variable.[7] Thus, an attempt was made to relate the pattern of price discrimination revealed by the country dummies (λ_i's) to market structure (concentration), rather than to product heterogeneity. It was found that exporter seller concentration did not significantly affect the export price. This result may reflect the sample used in the estimation.

The extended Knetter (1989) model was applied to data containing a cross-section of destination markets serviced during every time period. In the chicken export sample, these countries were generally the larger export markets and they generally had lower concentration ratios than markets not serviced in every period. One would not expect concentration to be significantly related to price in large markets serviced frequently by many firms. Recall, when using the export oligopoly model, the export price markup was found to be significantly related to export seller concentration. However, that model was estimated using a larger cross-section of markets, both large and small.

If the Knetter (1989) PTM model is generally estimated using data for large destination markets and if either government policy or marketing boards control imports, then the possibility of monopsonistic market power may be important. Pick and Park (1991) found that a variable measuring potential monopsony power had a significant affect on wheat export prices in their extension of the Knetter (1989) model.

Pricing to Market and Product Aggregation: U.S. Beef Exports

The Knetter (1989) PTM model was estimated for two beef products: (1) boneless, nonprocessed, fresh or chilled beef and (2) boneless, nonprocessed, frozen beef. These two products accounted for 21.4 and 27.1 percent of total U.S. beef export volume in 1990.[8] The model was also estimated for an aggregation of these products. Quarterly data for the years 1989 through 1991 were used, providing a total of 12 time periods for each destination. The same data sources used for chicken were also used for beef. These models were also estimated using ordinary least squares with restrictions on the time and country dummy coefficients.

Table 3.2 presents the results for the Knetter (1989) PTM model.[9] The null hypothesis that export prices to all destinations are equal is not rejected for either of the beef products (fresh or chilled or frozen). Only one country effect was found to be significantly different from zero. However, when these products were aggregated, it can be seen that the null hypothesis that all export prices are equal is rejected and that several country effects are significantly different from zero. This clearly illustrates the hazard of using aggregate data in this type of test.

It can be seen that the null hypothesis that all exchange rate coefficients equal zero is rejected at least at the ten percent level for the two beef products and the aggregate. However, for fresh or chilled beef shipments, the significant exchange rate coefficients (Japan, Singapore, and Hong Kong) take on a positive sign. This result raises the question of what the exchange rate coefficient truly measures. It may be the case that the exchange rate is correlated with other excluded variables that also affect export prices. The PTM hypothesis (negative β_j) is only supported in the case of frozen beef shipments to Sweden. When the aggregate product model is estimated, insignificant coefficients are found for some countries with significant exchange rate coefficients for the individual products. Thus, some of the exchange rate impacts are lost in the process of aggregation.

TABLE 3.2 Knetter Pricing to Market Model for U.S. Beef Exports

Country	(1) Fresh or Chilled Beef	(2) Frozen Beef	(1+2) Fresh, Chilled, or Frozen
Country Effects (λ)			
Japan	4.123* (1.756)	0.680 (1.302)	3.484** (2.919)
Korea		-0.037 (-0.028)	2.456 (0.799)
Taiwan	8.458 (0.981)		6.984 (0.895)
Singapore	-0.198 (-0.051)	-1.872 (-0.498)	0.349 (0.029)
Hong Kong	-12.383 (-1.506)	0.480 0.152	-19.234** (-2.161)
Sweden		0.750 (0.828)	5.961** (2.847)
F-Test ($H_0: \lambda_i = \lambda_j$)	1.663	0.504	3.624**
Exchange Rate Coefficients (β)			
Japan	1.47E-4** (2.533)	1.20E-5 (1.027)	6.47E-5* (2.314)
Korea		-0.329 (-0.533)	1.978 (1.359)
Taiwan	-0.397 (-0.200)		-0.241 (-0.145)
Singapore	3.851** (2.234)	0.131 (0.198)	0.830 (0.401)
Hong Kong	5.762** (2.234)	-0.267 (-0.393)	7.573** (2.808)
Sweden		-1.825** (-2.529)	0.745 (0.444)
F-Test ($H_0: \beta = 0$)	4.224**	2.118*	3.091**

Note: The values in parentheses are t-values. One and two asterisks (* and **) denote significance at the ten and five percent levels, respectively.

Summary and Conclusions

This chapter used the Knetter (1989) PTM model and an SCP type export oligopoly model to test for discriminatory pricing in export markets for U.S. chicken and beef products. In applying these alternative models to different products, some evidence on discriminatory export pricing behavior was found. Also, issues related to testing for discriminatory pricing in export markets were raised.

In the chicken export sector, the Knetter (1989) PTM model provided some evidence on discriminatory pricing by destination. Also, some evidence supporting the PTM hypothesis was found. However, it was argued that the evidence on discrimination by destination may be attributable to price differences due to product heterogeneity within the chicken commodity aggregate. In the analysis of U.S. beef exports, where more disaggregated data were available, it was shown that product heterogeneity may indicate the presence of price discrimination.

Some significant positive exchange rate coefficients were found using the Knetter (1989) PTM model for beef products. This indicates that U.S. beef exporters raised their export price for importers who experienced a currency depreciation. This counter-intuitive result raises the question on what the exchange rate coefficient truly measures. It is possible that the exchange rate is correlated with other excluded variables that may also affect export pricing. The relationship between exchange rates, other macroeconomic variables, and agricultural trade trends is now well known.

Using the export oligopoly model for U.S. chicken exports, it was shown that the U.S. exporter seller concentration and the importer's real national income were important in explaining the export price markup. Also, it was argued that potential monopsony power may be important in explaining export pricing behavior, particularly for the samples typically used in estimating Knetter's (1989) PTM model.

While both models can potentially provide evidence on export price discrimination, each has its own weaknesses. Many of the criticisms leveled against the traditional SCP models can be made of the export oligopoly model, as well. Some of the potential weaknesses of the Knetter (1989) PTM model were discussed here -- the model's failure to fully account for observed market structures on the export and import side of the transaction, the nature of competition among firms in these markets, and the uncertainty over what the estimated country dummies and exchange rate coefficients measure. Data aggregation, which is a problem using other models, was shown to be particularly problematic in the PTM model. At best, it might be said that these models provide a useful starting point. Future investigation should attempt to incorporate more fully important structural, strategic, and institutional characteristics of the market within defensible empirical frameworks.

Notes

1. Profitability is defined here as the extent to which price exceeds marginal cost.

2. Exchange rate pass-through measures the extent to which an importer's domestic currency import price from a particular source is changed due changes in the bilateral exchange rate. In a recent article, Knetter (1992) demonstrates how the concepts of PTM and exchange rate pass-through are related. When exchange rate pass-through is complete (the importer's price completely reflects the change in the exchange rate), PTM has not occurred. The lower the pass-through, the greater the degree of PTM.

3. It is recognized that cross-sectional heteroskedasticity and timewise autoregression may be a problem. However, the ordinary least squares estimates are still unbiased and consistent.

4. "National average" export price is defined as the mean of the country intercepts that are obtained from the model and is not truly the mean of the dependent variable.

5. Chicken is defined as an aggregate of the following U.S. Census export codes: 0207.10.4020, 0207.21.0020, 0207.39.0020, and 0207.41.0000.

6. The marginal cost of supplying each product in the aggregate is assumed to be constant across destinations.

7. $\ln p_{jt} = \theta_t + \lambda_j + \beta_j \ln s_{jt} + \gamma HHI_{jt} + v_{jt}$ $j=1,...,N$ $t=1,...,T$. Concentration was assumed to affect export pricing in the same manner in all markets. Thus, only a single concentration coefficient (γ) was estimated.

8. The beef products used here are for the U.S. Census export codes 0201.30.6000 and 0202.30.6000. Models for other beef products were not estimated, since there were not continuous shipments to a sufficiently large number of destinations throughout the sample period.

9. Concentration was also found not to have a significant effect on beef export prices.

References

Alston, J.M. and G.M. Scobie. 1987. "A Differentiated Goods Model of the Effects of European Policies in International Poultry Markets." *Southern Journal of Agricultural Economics* 19:59-68.

Baldwin, R. and H. Foster. 1986. "Marketing Bottlenecks and the Relationship Between Exchange Rates and Prices." MIT mimeo, Cambridge, MA.

Bessler, D.A. and J.A. Brandt. 1982. "Causality Tests in Livestock Markets." *American Journal of Agricultural Economics* 64:140-144.

Bresnahan, T.F. 1989. "Empirical Studies of Industries with Market Power," in R. Schmalensee and R. Willig, eds., *Handbook of Industrial Organization.* Amsterdam: North Holland.

Connor, J.M. and E.B. Peterson. 1992. "Market-Structure Determinants of National Brand-Private Label Price Differences of Manufactured Food Products." *Journal of Industrial Economics* 40:157-171.

Cowling, K. and M. Waterson. 1976. "Price Cost Margins and Market Structure." *Economica* 43:257-274.

Froot, K.A. and P.D. Klemperer. 1989. "Exchange Rate Pass-Through When Market Share Matters." *American Economic Review* 79:637-654.

Gundsman, C. and A. Webb. *Production, Supply, and Demand View.* U.S. Department of Agriculture, Economic Research Service, December 1991.

Hayes, D.J., J.R. Green, H.H. Jensen, and A. Erbach. 1991. "Measuring International Competitiveness in the Beef Sector." *Agribusiness, An International Journal* 7:357-374.

Helpman, E. and P.R. Krugman. 1989. *Trade Policy and Market Structure.* Cambridge, MA: MIT Press

Knetter, M.M. 1989. "Price Discrimination by U.S. and German Exporters." *American Economic Review* 79:198-210.

_____. 1992. "Exchange Rates and Corporate Pricing Strategies." NBER Working Paper Series, No. 4151.

Krugman, P.R. 1987. "Pricing to Market When the Exchange Rate Changes," in S.W. Arndt and J.D. Richardson, eds., *Real-Financial Linkages Among Open Economies.* Cambridge, MA: MIT Press.

Lyons, B. 1981. "Price-Cost Margins, Market Structure and International Trade," in D. Currie, D. Peel, and W. Peters, eds., *Microeconomic Analysis: Essays in Microeconomics and Economic Development.* London: Croom Helm.

Morgan, D. 1979. *Merchants of Grain.* New York, NY: Viking Press.

Pagoulatos, E. and R. Sorensen. 1976. "International Trade, International Investment and Industrial Profitability of U.S. Manufacturing." *Southern Economic Journal* 42:425-434.

Patterson, P.M. and P.C. Abbott. 1991. "An Evaluation of the PIERS Data for Use in Economic Analysis of U.S. Agricultural and Food Product Trade." NC-194 Occasional Paper OP-28. Columbus, OH: Ohio State University.

_____. 1992. "Further Evidence on Competition in U.S. Grain Export Trade." NC-194 Occasional Paper OP-35. Columbus, OH: Ohio State University.

Patterson, P.M., A. Reca, and P.C. Abbott. 1993. "Price Discrimination by U.S. High-Value Food Product Exporters: Empirical Evidence on U.S. Chicken, Beef, and French Fry Exports." Staff Paper No. 93-5, Department of Agricultural Economics. West Lafayette, IN: Purdue University.

Pick, D.H. and T.A. Park. 1991. "The Competitive Structure of U.S. Agricultural Exports." *American Journal of Agricultural Economics* 73:133-141.

Perez, A.M., M.R. Weimar, S. Cromer. 1991. *U.S. Egg and Poultry Statistical Series, 1960-90.* Statistical Bulletin 833. U.S. Department of Agriculture, Economic Research Service, Washington, D.C.

Pigou, A. C. *The Economics of Welfare.* London: McMillan, 1920.

Rogers, R.T. 1992. "Broilers - Differentiating a Commodity." Food Marketing Policy Center Research Report No. 18, Department of Agricultural and Resource Economics, Storrs, CT: The University of Connecticut.

Schmalensee, R. 1989. "Inter-Industry Studies of Structure and Performance," in R. Schmalensee and R. Willig, eds., *Handbook of Industrial Organization*. Amsterdam, North Holland.

Schroeter, J.R. 1988. "Estimating the Degree of Market Power in the Beef Packing Industry." *Review of Economics and Statistics* 70:158-162.

Suits, D.B. "Dummy Variables: Mechanics V. Interpretation." *Review of Economics and Statistics* 66:177-180.

Thornton, G. 1993. "Top 30 Export Markets for U.S. Broiler Meat." *Broiler Industry* 57(4):26-28.

4

Imperfect Competition and Exchange Rate Pass-Through in U.S. Wheat Exports

Timothy A. Park and Daniel H. Pick

Introduction

The general concept of imperfect competition has fascinated researchers over the years. Over the last two decades, we have seen a boom in the number of studies which have attempted to quantify this concept. The existing empirical studies have been directed at specific industries which were known to be characterized by high concentration ratios and lack of competition. These studies have attempted to evaluate the degree to which an industry exhibited monopoly power by measuring and estimating the markup over the estimated or observed marginal costs.

Beginning in the 1980s, we have also witnessed empirical applications of industrial organization models of imperfect competition to international trade. Besides theoretical attempts to show that large countries can gain by deviating from free trade, empirical studies have tried to determine whether countries which control large export shares exercise monopoly power in international markets.

Many of the early studies which addressed imperfectly competitive agricultural markets (McCalla 1966; Alaouze, Watson, and Sturgess 1978) have not explicitly tested the monopoly power hypothesis but rather justified using imperfectly competitive models in terms of large export shares. More recent studies in agricultural markets have tested the imperfectly competitive behavior hypothesis by using a variety of different approaches including the use of conjectural variations models (Thursby and Thursby 1990), estimating markup coefficients (Buschena and Perloff 1991; Love and Murniningtyas 1992), or testing for markup pricing across different markets (Pick and Park 1991).

Empirical testing of monopolistic behavior in international trade is both challenging and difficult. The standard test of pricing behavior, derived from basic microeconomic theory, is to observe whether price is equal to marginal cost. This approach requires data on prices and costs of production for a specific industry. While prices are usually observed, data on costs, especially in the context of international trade, are difficult and often impossible to find. One exception is the study by Buschena and Perloff (1991) on the Philippine coconut oil export market which used cost data to analyze market power behavior in the world market.

Many of the agricultural commodities which are traded internationally are dominated by a few large exporters or importers and are marketed through government agencies and marketing boards. These characteristics provide informal evidence which suggests that international agricultural markets are often characterized by imperfectly competitive behavior. Special attention has been directed at the international wheat market which is dominated by a few large exporters and importers.

Recently, a new concept has emerged which relates exchange rate changes to import prices. This new concept, titled "pricing to market," was introduced by Krugman (1987) and relates exchange rate pass-through to destination specific prices. Several studies have attempted to measure observed pass-through and its implications for market conduct (Feinberg 1986; Froot and Klemperer 1989; Knetter 1989). This approach is attractive since conclusions on market conduct can be drawn empirically without requiring data on costs of production.

Tests for pricing to market (PTM) behavior in agricultural trade have produced evidence of such conduct only in the wheat market (Pick and Park 1991). This chapter analyzes the international wheat market and provides further insights into the pricing to market behavior in that market. While previous studies have attempted to determine whether PTM exists, this chapter attempts to determine the sources of pass-through.

Just as in any model of market power, the starting point of this chapter is the equilibrium condition for profit maximization in noncompetitive behavior which equates marginal revenues to marginal costs in each destination market. A key factor to note is that marginal costs are common across destination markets, while markups over prices differ for shipments to each country.

However for a given destination market the adjustment of the optimal markup is symmetric with respect to changes in exchange rates and marginal costs. This symmetry condition will be derived in this chapter and can be imposed empirically in testing for pricing to market.

This chapter is organized as follows. The next section outlines the theoretical model used to justify the empirical application. The theoretical model is tested empirically for U.S. wheat exports in the next section. The hypothesis that exchange rate changes have similar effects on export prices is tested against the alternative hypothesis which states that these effects vary across destination markets. The last section presents a summary and conclusions.

Theoretical Model of Pricing to Market

The outline of the basic model to be estimated follows Knetter (1989, 1991) and is based on the export-pricing decisions of a multi-market monopolist. The firm exports to n different destination markets denoted by the index i. The profit equation is then given by:[1]

$$\pi(p_1, \cdots p_n) = \sum_{i=1}^{n} p_i q_i(e_i p_i) - C(\sum_{i=1}^{n} q_i(e_i p_i), w) \tag{1}$$

where p_i is the export price to market i denominated in the exporter's currency, q_i is the quantity demanded in market i and is a function of price denominated in the importer's currency. Thus e_i is the importer's currency per exporter's currency exchange rate. The firm's cost function is denoted by C, which is a function of quantity demanded and the input price w.

The first order condition for profit maximization implies that marginal revenue from shipments to each market is equated to the common marginal cost, or that price equals marginal cost plus an optimal markup in each destination market. This condition can be expressed in terms of the demand elasticity of imports:

$$p_i = C'(\frac{-\varepsilon_i}{-\varepsilon_i + 1}) \qquad \forall i \tag{2}$$

where C' is the derivative of the cost function with respect to q (the sum of exports over all destinations), and ε_i is the elasticity of demand in destination market i with respect to a change in price.

The impact of exchange rate changes on the export price is the sum of two distinct effects. The first effect measures the impact of exchange rate changes on marginal costs and is identical across destination markets.

The second effect highlights how shifts in exchange rates influence the import demand elasticity and the optimal markup. The latter effect is the exchange rate pass-through associated with PTM. The combination of these two effects is the concept referred to as exchange rate pass-through.

The above two effects are derived algebraically in Knetter (1991) by converting equation (2) into logarithmic form. By totally differentiating (2) with respect to input prices, export prices and exchange rates, the following expression is obtained:

$$\frac{dp_i}{p_i} = (1+\beta_i)\frac{dC'}{C'} + \beta_i\frac{de_i}{e_i} \quad \forall i, \text{ where}$$

$$\beta_i = \left(\frac{\frac{\partial\ln\varepsilon_i}{\partial\ln(e_ip_i)}}{(-\varepsilon_i+1)-\frac{\partial\ln\varepsilon_i}{\partial\ln(e_ip_i)}} \right) \tag{3}$$

Equation (3) demonstrates the symmetric effect that changes in marginal costs and changes in exchange rates have on prices. In fact, if equation (3) was stated such that prices were denominated in the importer's currency, these effects would be identical. Symmetric pass-through effects of tariff and exchange rates on prices have been analyzed for the auto industry by Feenstra (1989) and for Brazilian tobacco trade by Pompelli and Pick (1990).

The model of a multi-market monopolist is a simplification that can be modified to apply to alternative specifications of market power. Knetter (1991) suggested that when exporters face competition from other countries, the optimal markup across destinations is based on the residual demand curve perceived by the exporting country.

Feenstra (1989) also developed a symmetry hypothesis for pass-through effects of tariffs and exchange rates based on a general model of imperfect competition. He noted that similar implications are obtained when firms behave as either Bertrand or Cournot competitors, indicating that the general results do not depend on assumptions about market structure.

Empirical Specification

The empirical model based on the theoretical restrictions developed in equation (3) requires information on marginal cost over time along with exchange rates for each destination country. The econometric model based on equation (3) is:

$$\ln p_{it} = (1 + \gamma_i)\ln c_t + \beta_i \ln e_{it} \qquad\qquad i = 1,...,N \quad t = 1,...,T \qquad (4)$$

where i indicates the N destination markets and t represents the T quarterly time periods.

A major difficulty in estimating this model is the lack of data on the marginal cost of producing wheat across the different time periods. To overcome this difficulty, quarterly time dummy variables, θ_t, capture the common marginal cost across destination markets. The elasticity of export price with respect to exchange rate for each destination country is $(1 + \beta_i)$. The time dummy variables will also reflect common factors such as shifts in demand or factors that induce changes in industry conduct which influence export prices to each destination.

Knetter (1989) proposed a model to test the symmetry restriction of marginal cost and exchange rate pass-through by exploiting the information on shipments to destination markets over time. For the producer, the level of marginal cost for exporting wheat to each destination market is identical. Marginal costs may shift over time but will not vary across destination markets. In this model, shifts in marginal costs over time can be captured by time-specific effects.

The econometric model with the symmetry restriction imposed can be then specified as:

$$\ln p_{it} = (1 + \beta_i)\theta_t + \beta_i \ln e_{it} + u_{it} \qquad\qquad (5)$$

where θ_t is a time-specific effect and the error term u_{it} is identically and independently distributed.

An alternative model which relaxes the symmetry restriction compares the validity of the symmetry restriction implied by pricing to market behavior. This model is specified as:

$$\ln p_{it} = \theta_t + \beta_i \ln e_{it} + u_{it} \qquad\qquad (6)$$

where θ_t the time-specific effect measures variables which affect export prices proportionately across all destinations.

Arguments are suggested to support the alternative model which relaxes the symmetry restriction. First, a set of common factors besides changes in marginal cost could induce exporters to adjust prices proportionately across destinations. Any shocks to demand for the exported good that are shared across markets, along with shifts in industry pricing and marketing practices that are common to each destination, may lead to a rejection of the symmetry model.

Second, exporters may have better information for forecasting changes in marginal cost than in predicting changes in exchange rates. Asymmetric adjustment patterns in marginal costs imply that changes in marginal cost are more permanent than changes in exchange rates. As a result, export pricing decisions may adjust more quickly to changes in marginal cost than to shifts in exchange rates.

The pricing equations (5) and (6) are jointly estimated for each of the U.S. wheat export destination markets. The linear model in equation (6) is a seemingly unrelated regression in which the time-specific effects, θ_t, are restricted to be equal for each destination. In the nonlinear model defined in equation (5), the model is estimated imposing the additional equality restrictions that the β coefficients for the same destination are equal across time periods.

The model in equation (5) is applied to U.S. wheat exports to test for noncompetitive market conduct by the United States. The commodity group is wheat not donated (0410040). Data were compiled on a quarterly basis for the 1978-1991 period. Quantity and value data were available from the U.S. Department of Commerce Schedule E. The value data are free alongside ship (FAS), which exclude the cost of loading or any other charges or any transportation costs beyond the port of destination. The quantity and value data were used to generate the price (unit value) variable.

Exchange rate data were available from the *International Financial Statistics* published by the International Monetary Fund. Real exchange rates were calculated using the consumer price index (CPI) for each country. Official exchange rates for the Soviet Union and the People's Republic of China were used in the study.[2]

The error terms in equations (5) and (6) may be linked to measurement error in the dependent variable. The unit value data were generated using quantity and value data in place of observed transaction prices. The empirical models are valid if the measurement errors for each destination are uncorrelated with the exchange rate for that destination. The time-specific effects will capture any changes in quality and price which are common to all destinations over the time period.

Another source of measurement error is associated with structural change in the β_i coefficients over time which may cause the coefficients to shift. An example of such structural change is the implementation of the Export Enhancement Program (EEP) during the period used in the analysis.

The program was established in response to export subsidies by the European Community. Under the program, targeted countries are eligible for subsidized wheat exports. Four of the destination markets used in this study participated in the Export Enhancement Program

(EEP): Egypt, the People's Republic of China, Philippines, and the former Soviet Union. This structural change was accounted for in the model by incorporating an EEP dummy variable for the participating countries in the corresponding time periods.

A final point to note is that the symmetry restriction is derived from a comparative static model which abstracts from short term deviations from market equilibrium. The logarithmic specification of the econometric models adopted in this chapter assumes that any movements away from equilibrium are transitory and are represented by a random error term.

Alternatively, the error term may represent the impact of stochastic shocks which result in permanent shifts to the equilibrium relationship between export prices, costs, and exchange rates. These permanent shifts imply the existence of a model with a unit root in the error term, commonly referred to as a "difference-stationary" (DS) model.

The presence of a unit root in the error term of the logarithmic specification is rejected, establishing support for empirical models to test the symmetry restriction. The Dickey-Fuller test rejects the null hypothesis of a unit root in the residuals of the logarithmic specification. The test statistic of -13.619 is compared to the asymptotic critical value of -3.66 at the 2.5 percent level. Critical values are taken from Table 20.1 presented in Davidson and MacKinnon (1993).

The empirical results are based on a logarithmic specification for equations (5) and (6). This specification is justified given that the presence of a unit root is rejected in the error term of the logarithmic model.

Results of the Estimated Models

The results for the symmetry model and the unconstrained model are presented in Tables 4.1 and 4.2, respectively. For each destination the elasticity of markups with respect to the normalized real exchange rates is measured by the β_i coefficients.

Coefficients for the β_i's that are significantly different from zero imply behavior which is consistent with pricing to market across destination markets. Negative β_i coefficients suggest a model of price discrimination in which exporting firms adjust prices in export markets to offset exchange rate movements in each destination market. The term associated with the time-specific effect, θ_t, measures exchange rate pass-through net of changes in marginal cost.

Comparison of the results of the constrained (symmetric) model in Table 4.1 with those of the unrestricted model in Table 4.2 reveal some differences. The hypothesis of PTM behavior in the international wheat market is strongly supported by the symmetric model.

The econometric model and the symmetry conditions are derived directly from the profit maximization first-order conditions. Given that PTM behavior has been observed in the symmetric model as well as in previous studies (Pick and Park 1991; Pick and Carter 1993), we focus on the interpretation of results from the symmetry model.

TABLE 4.1 Pricing to Market Model for U.S. Wheat Exports -- Symmetry Equation

Coefficient Symbols	Country	Parameter Estimates	Standard Error
β_1	People's Republic of China	-0.9197	0.9865
β_2	Egypt	-0.9956*	0.0551
β_3	Japan	-0.9861*	0.1717
β_4	Korea	-0.9852*	0.1817
β_5	Philippines	-0.9988*	0.0156
β_6	Soviet Union	-0.8173	2.2376
β_7	Taiwan	-0.9851*	0.1840
β_8	Venezuela	-0.9797*	0.2487
E_1	EEP Dummy PRC	-2.2783	1.4789
E_2	EEP Dummy Egypt	0.7820*	0.2927
E_5	EEP Dummy Philippines	1.3654*	0.2677
E_6	EEP Dummy Soviet Union	-8.5614	2.0896

Note: * indicates t-values significant at the 5 percent level.

TABLE 4.2 Pricing to Market Model for U.S. Wheat Exports
-- Unrestricted Model

Coefficient Symbols	Country	Parameter Estimates	Standard Error
β_1	People's Republic of China	-4.5043	1.0750
β_2	Egypt	-0.4507	0.8970
β_3	Japan	-0.3111	1.4810
β_4	Korea	2.0980	2.2321
β_5	Philippines	0.4235	2.8922
β_6	Soviet Union	9.4962	3.5328
β_7	Taiwan	1.5088	2.6089
β_8	Venezuela	0.0828	0.9255
E_1	EEP Dummy - PRC	4.2653*	0.9620
E_2	EEP Dummy - Egypt	0.0498	0.6562
E_5	EEP Dummy - Philippines	0.1404	0.8401
E_6	EEP Dummy - Soviet Union	-1.9672*	0.8425

Note: * indicates t-values significant at the 5 percent level.

Results from the symmetric model in Table 4.1 indicate that in six of the eight countries in the sample, the β_i coefficients are negative and significant: Egypt, Japan, Korea, Philippines, Taiwan, and Venezuela. For two of the countries, the People's Republic of China and the Soviet Union, the β_i coefficients were negative but not significant.

The fact that these coefficients are insignificant is not surprising given the nature of the exchange rate data used in the analysis for these countries. This point is noted in the discussion on the exchange rates used for the People's Republic of China and the Soviet Union. In fact, it would have been more surprising to observe significant coefficients for these countries.

In the model proposed by Knetter (1989), exchange rate pass-through net of changes in marginal cost is measured by the term $(1 + \beta_i)$. The β value for Japan of was -0.9861. The markup of U.S. export prices over marginal cost will fall by 1.39 percent in response to a 10 percent appreciation of the U.S. exchange rate relative to the Japanese yen. For constant marginal costs this implies that exchange rate pass-through is

13.9 percent. Seven of the eight destination-specific markups are below 2 percent, with the Soviet Union the largest at 18 percent.

The differences in the estimated coefficients between the symmetric model and the unconstrained model indicate some instability across the specifications. In the unconstrained model the β_i coefficients are negative for three countries: the People's Republic of China, Egypt, and Japan. The coefficients on the remaining countries are positive but statistically insignificant, with the exception of the former Soviet Union. As Knetter (1991) noted a positive markup suggests that price adjustments across destination markets by exporting firms work to exacerbate the impact of exchange rate movements.

In the unconstrained model the time dummy variables are assumed to account for all common movements in export prices such as shifts in demand and changes in industry marketing behavior across all destinations. However, common factors that influence the marketing behavior of all exporting countries may play a limited role in the world wheat market. Wilson, Koo and Carter (1990) noted that exporters have made efforts to differentiate the quality and characteristics of their wheat exports to maintain market shares. If international trade in wheat is differentiated across markets then there may not exist a set of strong common factors that affect export prices proportionately across each destination market. The validity of the alternative model which relaxes the symmetry restriction is in turned weakened.

Feenstra (1989) also noted that there is no good alternative hypothesis to the symmetry model but cautioned that alternative specifications should be considered in estimating the price adjustment equation.

Conclusions

Exchange rate pass-through and PTM have often been discussed jointly when in fact these are two different concepts. While exchange rate pass-through is associated with changes in prices associated with changes in marginal costs as the exchange rate changes, PTM is a term associated with imperfectly competitive behavior. Changes in prices which are associated with changes in exchange rates are related to the convexity of the demand schedules.

In this chapter, we estimate a model which allows us to test for imperfectly competitive behavior associated with changes in exchange rates. The strength of this approach is that it can incorporate the impact of changes in marginal costs by using time dummy effects across destinations even without observing actual cost data.

The estimated model is based on a symmetry restriction derived from the export-pricing decisions of a multi-market monopolist. The results suggest the existence of PTM behavior in the international wheat market as six of the eight exchange rate coefficients were significant and negative. The nonlinear version of the model imposes a symmetry restriction between the effects of marginal costs and exchange rates on export prices. This symmetry condition is a direct implication derived from the model of profit maximization. The results also confirm the significant effects of the Export Enhancement Program on destination specific wheat prices.

Notes

1. In this equation we ignore any domestic sales component.
2. The shortcomings of these exchange rates are well acknowledged. The exchange rate coefficients for these countries in Table 4.1 were not significant.

References

Alaouze, C.M., A.S. Watson, and N.H. Sturgess. 1978. "Oligopoly Pricing in the World Wheat Market." *American Journal of Agricultural Economics* 60:173-85.

Buschena, D. and J.M. Perloff. 1991. "The Creation of Dominant Firm Market Power in the Coconut Oil Export Market." *American Journal of Agricultural Economics* 73:1000-1008.

Davidson, R. and J.G. MacKinnon. 1993. *Estimation and Inference in Econometrics.* New York, NY: Oxford University Press.

Feenstra, R.C. 1989. "Symmetric Pass-Through of Tariffs and Exchange Rates Under Imperfect Competition: An Empirical Test." *Journal of International Economics* 27:25-45.

Feinberg, R.M. 1986. "The Interaction of Foreign Exchange and Market Power Effects on German Domestic Prices." *Journal of Industrial Economics* 35:61-70.

Froot, K. and P. Klemperer. 1989. "Exchange Rate Pass-Through When Market Share Matters." *American Economic Review* 79:637-654.

Knetter, M.M. 1989. "Price Discrimination by U.S. and German Exporters." *American Economic Review* 79:198-210.

_____. 1991. "Pricing to Market in Response to Unobservable and Observable Shocks." Dartmouth College, Department of Economics, Working Paper No. 89-16.

Krugman, P.R. 1987. "Pricing to Market When the Exchange Rate Changes," in S.W. Arndt and J.D. Richardson, eds., *Real-Financial Linkages Among Open Economies.* Cambridge, MA: MIT Press.

Love, A.H. and E. Murniningtyas. 1992. "Japanese Wheat Imports: A Test of Market Power." *American Journal of Agricultural Economics* 74:546-555.

McCalla, A.F. 1966. "A Duopoly Model of World Wheat Pricing." *Journal of Farm Economics* 48:711-727.

Pick, D.H. and C.A. Carter. 1993. " Pricing to Market with Transactions Denominatd in a Common Currency." *American Journal of Agricultural Economics* 76:55-60.

Pick, D.H. and T.A. Park. 1991. "The Competitive Structure of U.S. Agricultural Exports." *American Journal of Agricultural Economics* 73:134-141.

Pompelli, G.K. and D.H. Pick. 1990. "Pass-through of Exchange Rates and Tariffs in Brazil-U.S. Tobacco Trade." *American Journal of Agricultural Economics* 72:676-681.

Thursby, M.C. and J.G. Thursby. 1990. "Strategic Trade Theory and Agricultural Markets: An Application to Canadian and U.S. Wheat Exports to Japan," in C. Carter, A.F. McCalla and J. Sharples, eds., *Imperfect Competition and Political Economy: The New Trade Theory in Agricultural Trade Research.* Boulder, CO: Westview Press.

Wilson, W. W. Koo, and C.A. Carter. 1990. "Importer Loyalty in the International Wheat Market." *Journal of Agricultural Economics* 41:94-102.

5

The Political Market for Protection in U.S. Food Manufacturing

Rigoberto A. Lopez and Emilio Pagoulatos

Introduction

The food processing industries of the United States are characterized by trade barriers that are more restrictive than most manufacturing industries (Lavergne 1983; Ray 1990). These trade barriers, which include nominal tariffs and nontariff barriers (NTBs), act as entry barriers from the standpoint of foreign competitors. The strategic implication is that firms will spend resources to compete in the political arena to influence trade barriers in their favor, instead of concentrating their efforts on regular market competition, such as pricing and other nonprice strategies. As a result, a significant redistribution of wealth may occur as trade barriers increase prices and food processing industry profits at the expense of domestic consumers. In addition, resources are spent on unproductive activities, such as lobbying, which may result in a decline in overall economic efficiency (Lopez and Pagoulatos 1994).

The intent of this study is to explain the variation in the level of trade barriers (tariffs and import-quota tariff equivalents) across U.S. food processing industries in terms of strategic behavior of firms as well as industry structure characteristics. The general framework of analysis is the theory of public choice, which analyzes political processes and their interaction with the economy (Mueller 1979). Of particular interest is the importance of interest group pressures (lobbying) in determining the level of trade barriers. In addition to addressing trade barriers pertinent to food processing, this chapter incorporates more direct measures of lobbying activity than previous studies of manufacturing industries, such as the contributions of Political Action Committees (PACs) connected to the respective industries at the 4-digit SIC level.

The first section of this chapter presents the conceptual framework for analyzing the economics and politics of trade barrier formation. The second section discusses the data and econometric specification, and the third section analyzes the empirical results. Brief concluding remarks appear in the final section.

Conceptual Framework

A widely accepted and useful framework to analyze the determinants of trade barriers is one that conceptualizes them within a political market for protection (Anderson and Baldwin 1981; Nelson 1988; Baldwin 1989; Ray 1981a, b, 1990). Therefore, domestic food manufacturers can be viewed as the "demanders" and government legislators and bureaucrats as "suppliers" of trade barrier protection.

The theory of rent-seeking (recently surveyed by Tollison 1982, and Brooks and Heijdra 1989) provides the arguments for the demand side of protection. Tariffs and import quotas artificially increase the price of commodities while restricting foreign entry, hence either generating rents or avoiding or slowing the demise of a domestic industry. For an industry to exert political pressure, a necessary condition is that it be willing and able to organize its members to obtain contributions needed for lobbying.

Past studies have used several industry characteristics as proxies for political activity that will presumably result in political outcomes such as tariffs (e.g., Caves 1976; Saunders 1980; Ray 1981a, b). Since trade policy can be viewed as a public good, a common-interest group advocating such a policy needs to overcome the free rider problem typically present in raising lobbying funds. Thus, lobbying activity is usually approximated by industry organization characteristics, such as the number of firms, the concentration ratio, and the size of the industry that determine the degree of free riding. More recently, Baldwin (1989), among other economists, has advocated the use of more direct measures, such as funding levels and number of employees supported by private interests, for lobbying purposes. An important motivation for demanding food manufacturing protection may be the ultimate protection of the underlying agricultural industry. Protection at the food processing level may be needed in order to protect farm policy programs because the farm industry is often integrated forward into processing. Examples of these two situations involve the U.S. sugar (cane and beet) and milk industries.

The supply side of the protection market is usually characterized either by political behavior concerned with re-election, i.e. the maximization of votes (Peltzman 1976; Becker 1983), or concerned with social welfare (principled behavior). If behavior is characterized by vote-seeking, then policymakers or bureaucrats will make choices to balance political support from conflicting interest groups. Thus, to some extent, they will take into account the interests of consumers and will be concerned with public sympathy or dislike for government protection through trade barriers. If behavior is driven by societal principles, then policymakers will make choices that reflect the social consensus on income distribution, employment, and social goals in general. Both types of behavior, alternately called passive and self-willed governments by Bhagwati (1989), are likely to coexist in the case of food manufacturing industries.

An important element of the supply side of the market for protection is the degree of comparative advantage. First, import trade barriers are redundant or ineffective if an industry has comparative advantage in trade. Second, industry lobbying is more effective in gaining the sympathy of policymakers and the public if an industry has some degree of comparative disadvantage. Comparative advantage is usually measured by (1) the growth in employment, output or exports, (2) the capital/labor ratio or the degree of labor intensity in production, and (3) the skill level of the labor force (Anderson and Baldwin 1981; Lavergne 1983; Ray 1981a, b, 1990). If an industry is declining, then that industry will more likely be protected by a higher tariff level or a more restrictive import quota. Traditionally, U.S. comparative disadvantage has been established in low-skilled, labor-intensive industries.

Based on the above discussion, we can broadly hypothesize that trade barrier formation in the food processing industries can be summarized by the following function:

$$T_k = f\ (Z_k, Z_{ok}, CD_k,\ O_k)$$

where T_k is the tariff level or tariff equivalent of an import quota in the kth industry[1], Z_k is the lobbying activity by that industry, Z_{ok} is the lobbying activity by those adversely affected by the tariff or quota, CD_k is a vector of variables describing the degree (if any) of comparative disadvantage, and O_k is a vector of other variables, such as industry structure characteristics.

Empirical Model Specification

The statistical objective of our data analysis is to determine the influence of economic and political variables on the inter-industry pattern of trade barriers in U.S. food and tobacco processing. The dependent variable (T) used in this study was the simple average U.S. nominal tariff rate based on the c.i.f. value of imports at the 4-digit SIC level for 1987. The tariff rates were obtained from a computer tape supplied by the U.S. International Trade Commission [USITC]. Tariff equivalents were used in those industries where import quotas were the main instrument of protection. Data on tariffs equivalents were obtained from International Trade Commission reports (USITC 1990a, 1990b).[2]

Several independent variables were included to account for the major hypothesized factors discussed in the previous section. Two variables commonly used to represent market structure were included: the four-firm concentration ratio (CR) and the number of companies (N). The impact of market structure characteristics on trade barrier restrictions is somewhat ambiguous. For instance, Ray (1981a, b) argues that the impact of seller concentration on trade restrictions is ambiguous. From the industry perspective, the greater the seller concentration, the more effective is the organization for lobbying, and the greater the per capita gains. From the policymakers' perspective, the spoils of protection will buy more votes the more widely the industry is dispersed (Caves 1976). Regarding the number of firms, Godek (1985) argues that it should be negatively related to trade protection because the more firms in the industry, the more costly its organization (free rider problem) and the lower the per firm transfers. However, as the number of firms expands, the number of beneficiaries from protection increases, possibly increasing the political pay-off. As noted by Peltzman (1976), the size of the transfer depends on support, opposition, and organizational cost. Thus, N^2 was included to capture any nonlinearities of the impact of N on trade protection decisions. Both variables, CR and N, came from the 1987 Census of Manufactures.

To control for comparative advantage, we included four variables: value-added per employee (VAE); export share (XS); foreign direct investment inflow (FDI); and growth in employment (GE).

The variable VAE, a proxy for total (human and physical) capital per person, was computed from the 1987 Census of Manufacturers. The variable should be negatively related to trade protection to the extent that it directly measures comparative advantage (Ray 1981a, b; Godek 1985). The variable XS was computed as the ratio of F.A.S. value of exports to apparent consumption. This variable should be negatively related to trade protection to the extent that a higher export share is a reflection of domestic comparative advantage relative to foreign producers. The FDI variable was measured as a ratio of foreign direct investment inflows to total domestic assets and obtained from Pagoulatos (1983) and MacDonald and Weimer (1983). Because this variable suggests the presence of intra-firm trade, an increase in this ratio should reduce an industry's benefits from additional protection. The growth in employment variable was defined as the change in employment from 1977 to 1987 (1987 Census of Manufactures). This variable should be inversely related to trade barriers, as trade protection is expected to be higher in declining industries.

Four direct political variables were included: campaign contributions (PAC), agricultural interests (AGINT), consumer interests (CONS), and structure of Congress (SCONG). The variable PAC was measured by the total dollars contributed to the 1987-88 congressional elections by each industry divided by the industry's value added (PAC). The PAC funding data came from Makinson (1990), while the value added data came from the 1987 Census of Manufactures. A positive impact of PAC on trade protection is expected as it results in a higher demand for protection (Abler 1991). Agricultural interests were measured by the share of agriculture in the value of manufactured food product in the pertinent SIC industry, obtained from the 1977 input-output tables of the U.S. Lobbying by the underlying agricultural establishment is taken as complementary to food processing, and hence AGINT is expected to have a positive impact on trade barriers. The consumers good dummy, by indicating that the product in question does not represent a major input cost to other manufacturers, should be positively related to the level of protection. This is because interests of consumers of final goods are likely to be more dispersed than those of buyers of intermediate goods (Ray 1981a; Godek 1985). Finally, the structure of Congress was measured by adding the number of Congresspersons from states producing over five percent of the total value of shipments. This variable is expected to positively affect the level of trade protection due to the fact that a greater value of SCONG denotes wider congressional representation for a specific industry.

The equation which summarizes the empirical determinants of the level of trade barriers in food processing industries discussed above is:

$$T = f\ (CR, N, N^2, GE, FDI, XS, VAE, PAC, AGINT, CON, SCONG)$$
$$\pm\quad -\ +\quad -\quad -\quad -\quad -\quad +\quad +\quad +\quad +$$

The sign below a variable indicates the hypothesized sign of its coefficient; a ($\pm$) indicates that no *a priori* sign prediction is warranted. The model coefficients were estimated using data for forty four 4-digit SIC food and tobacco manufacturing industries for the year 1987 via ordinary least squares. Three versions of the model were estimated to assess the importance of the market structure and comparative disadvantage vis-á-vis pure political variables: a model that included the model which included all variables (Model 3). The results are presented below.

Empirical Results

The empirical results are presented in Table 5.1. To investigate the relevance of the alternative models, we conducted F-tests of the hypothesis that all the coefficients in each model were zero. This hypothesis was rejected for Models 2 and 3 as the resultant F-statistics exceed the critical F-values at the 5 percent level. However, for Model 1, we fail to reject the hypothesis that all the coefficients are significantly different from zero, as the resulting F-statistic (1.274) with (7,35) degrees of freedom did not exceed the critical F-value of 2.29 at the 5 percent level. This implies that the sole use of comparative disadvantage and industry structure variables is unsatisfactory in modeling the determinants of U.S. food processing trade barriers and supports the use of direct political variables.

To explore further issues of model specification, F-tests were conducted to test whether Model 3 added significant amounts of information over the restricted versions given by Models 1 and 2. The F-statistic for Model 1 versus Model 3 was 4.242 with (7,32) degrees of freedom which exceeds the critical F-value at the 5 percent (critical F=2.31) and at the 1 percent (critical F=3.25) levels. This indicates that Model 2 (the omitted variables from Model 1) add a significant amount of information. The F-statistic for Models 2 versus 3 (3.109 with (4,32) degrees of freedom) was also significant at the 5 percent level but not at the 1 percent level (critical F=3.99). Thus, these results give some support to including the comparative disadvantage and industry structure variables in addition to the political influence ones. Except for Model 1, the models were quite satisfactory in explaining trade protection in food and tobacco manufacturing in 1987, especially for a small, cross-sectional sample. The adjusted R^2 for Models 2 and 3 were .399 and .472 (Anderson and Baldwin 1981 and Ray 1981a, b, obtained smaller R^2s for much larger samples).

Given the results from the model specification tests, we focus on the results from the full model. First, we turn to the results for the market structure variables: CR, N, and N^2. All three variables were significant at least at the 10 percent level. The results indicate that the level of protection is inversely related to seller concentration. This result supports the hypothesis that trade protection tends to be lower in highly concentrated industries. As hypothesized, the number of firms exhibited a nonlinear relationship with the level of protection. As the number of

TABLE 5.1 Determinants of Trade Protection in U.S. Food Manufacturing, 1987

Variable	Model 1	Model 2	Model 3
Four-Firm Concentration Ratio (CR)	-.975** (2.185)		-.9772*** (2.865)
No. of Companies (N)	-.001** (2.034)		-.0006* (1.930)
No. Companies squared (N^2)	3.3×10^{-7} (1.634)		$3 \times 10^{-7*}$ (1.658)
Employment Growth 1977-87 (GE)	.184 (.771)		-.066 (.358)
Value Added per Employee (VAE)	.001 (.936)		.002** (2.317)
Export Share (XS)	-.699* (1.669)		-.630* (1.900)
Foreign Direct Investment (FDI)	-.016* (1.702)		-.017** (2.291)
PAC Intensity (PAC)		1.932*** (4.742)	1.732*** (4.322)
Agricultural Interests (AGINT)		.389** (2.229)	.456*** (2.504)
Consumer Product (CONS)		0.264*** (2.556)	0.323*** (3.073)
Structure of Congress (SCONG)		.002* (1.817)	.001 (1.104)
Constant	.736** (2.162)	-.509*** (2.671)	.173 (.549)
R^2 Adjusted	.043	.399	.472
F-Ratio	1.274	8.125	4.497
Sum of Squared Errors	4.242	2.887	2.079

Note: One, two and three asterisks (*) indicate statistical significance at the 10, 5 and 1 percent levels. The absolute values of the t-statistics are given in parentheses.

firms increases, there is a downward pressure on trade protection, possibly as a result of increased organizational cost for lobbying. However, due to the nonlinear impact of N, a threshold point is reached ($N^* = 1.299$) beyond which an increase in N results in a higher level of protection. Our results are consistent with both Peltzman's (1976) and Godek's (1985) arguments with respect to the number of beneficiaries.

Turning next to the results for the comparative advantage variables, only the employment growth variable was not significantly different from zero at the 5 or 10 percent levels. Thus, the model and the data fail to support the hypothesis that trade protection responds to losses in employment in the food industries. However, the sign of the coefficient associated with value-added per employee (VAE) is contrary to expectations, suggesting that in U.S. food manufacturing the structure of nominal protection is favorable to industries with high value added per person. As hypothesized, export oriented industries tend to face lower levels of trade protection. The results clearly show that foreign direct investment (FDI) inflows in the U.S. food manufacturing industries are associated with lower levels of protection. As foreign direct production and marketing occur within the borders, traditional exporting from abroad and hence trade protection become less imperative.

Finally, except for the structure of Congress, all the direct political variables were significant at the 5 percent level. Thus, the spatial structure of production, as a proxy for congressional representation, fails to show any measurable impact on trade protection decisions. As expected, PAC contributions strongly influence the level of trade protection. It is estimated that a 1 percent increase in PAC contribution intensity (PAC contributions as percentage of value added) leads to nearly 5 percent increase in tariff rates or their equivalent. The results show that agricultural interests are associated with higher trade protection in food manufacturing. A 1 percent increase in the share of agricultural value in the final food product leads to approximately 0.46 percent increase in the trade protection rate. The results also show that consumer products do bear higher levels of trade protection, consistent with the hypothesis that such decisions tend to find less opposition than trade protection on intermediate products, as found by Ray (1981a) and Godek (1985).

Conclusion

This study examined the determinants of the inter-industry pattern of trade barriers in forty-four food and tobacco manufacturing industries in 1987. Although traditional market structure and comparative advantage variables are important determinants of the level of trade protection, this chapter clearly underscores the importance of including direct political variables in models explaining the forces behind observed protection levels in U.S. food manufacturing industries. More specifically, trade protection decreases with seller concentration and exhibits a nonlinear relationship with respect to the number of firms. In addition, the empirical results provide some support to the hypothesis that trade barriers are higher in those industries in which the United States has a long standing comparative disadvantage.

Where our results break new ground is in the importance of including direct lobbying intensity variables as explanations of the level of trade protection in U.S. food industries. In particular, the amount of contributions by political action committees as a percentage of value added and agricultural interests as well as those of other manufacturers are the most significant variables in determining the interindustry variation of tariffs and tariff-equivalents of quotas in U.S. food and tobacco processing.

Notes

1. The efficiency equivalence between tariffs and tariffs-equivalents of import quotas (those tariffs that result in the same level of imports as the quota) has been questioned due to such dynamic considerations as market power (Bhagwati 1965). In terms of political inequivalence, much of the empirical evidence supports the notion that quotas are used to pursue higher levels of protection because of their informational advantages (Godek 1985), their transparency in trade negotiations (Ray 1981a, b), their ability to overcome free riding and to be less visible (Kaempfer and Willet 1989), or their expediency in terms of hidden consumer cost and direct tax or treasury expenditure avoidance (Lopez 1989).

2. Food industries protected by quotas rather than tariffs include meat packing (SITC = 2011), creamery butter (2021), cheese (2022), condensed and evaporated milk (2023), fluid milk (2026), and refined sugar (2061-3).

References

Abler, D.G. 1991. "Campaign Contributions and House Voting on Sugar and Dairy Legislation." *American Journal of Agricultural Economics* 73:11-17.

Anderson, K. and R.E. Baldwin. 1981. *The Political Market for Protection in Industrial Countries: Empirical Evidence.* Washington D.C.: World Bank Staff Working Paper No. 492.

Baldwin, R. E. 1985. *The Political Economy of U.S. Import Policy.* Cambridge, MA: MIT Press.

_____. 1989. "The Political Economy of Trade Policy." *Journal of Economic Perspectives* 3:119-135.

Becker, G.S. 1983. "A Theory of Competition Among Pressure Groups for Political Influence." *Quarterly Journal of Economics* 98:371-400.

Bhagwati, J.N. 1965. "On the Equivalence of Tariffs and Quotas," in R. E. Baldwin et al., eds., *Trade, Growth and the Balance of Payments*. Chicago, IL: Rand McNally.

_____. 1989. "Is Free Trade Passé After All?" *Weltwirtschaftliches Archiv* 125, 1:1744.

Brooks, M.A. and B.J. Heijdra, 1989. "An Exploration of Rent Seeking." *Economic Record* 65:32-50.

Caves, R.E. 1976. "Economic Models of Political Choice: Canada's Tarrif's Structure." *Canadian Journal of Economics* 9:278-300.

Godek, P.E. 1985. "Industry Structure and Redistribution through Trade Restrictions." *Journal of Law and Economics* 28:687-703.

Kaempfer, W.H. and T.D. Willet. 1989. "Combining Rent-Seeking and Public Choice Theory in the Analysis of Tariffs versus Quotas." *Public Choice* 59:79-86.

Lavergne, R.P. 1983. *The Political Economy of U.S. Tariffs: An Empirical Analysis*. New York: Academic Press.

Lopez, R.A. 1989. "Political Economy of U.S. Sugar Policies." *American Journal of Agricultural Economics* 71:20-31.

Lopez, R.A. and E. Pagoulatos. 1994. "Rent Seeking and the Welfare Cost of Trade Barriers." *Public Choice* 79:149-160.

MacDonald, J.M. and S.A. Weimer. 1985. *Increase Foreign Investment in U.S. Food Industries*. Economic Research Service, Agricultural Economics Report No. 540, Washington, D.C.: U.S. Department of Agriculture.

Makinson, L. 1990. *Open Secrets: The Dollar Power of PACs in Congress*. Center for Responsive Politics, Congressional Quarterly Inc. Washington, D.C.:

McKenzie, R.B. 1988. "The Relative Restrictiveness of Tariffs and Quotas: A Reinterpretation from a Rent-Seeking Perspective." *Public Choice* 58:85-90.

Mueller, D.C. 1979. *Public Choice*. Cambridge: Cambridge University Press.

Nelson, D. 1988. "Endogenous Tariff Theory: A Critical Survey." *American Journal Political Science* 32:796-837.

Pagoulatos, E. 1983. "Foreign Direct Investment in U.S. Food and Tobacco Manufacturing and Domestic Economic Performance." *American Journal of Agricultural Economics* 65:405-412.

Peltzman, S. 1976. "Toward a More General Theory of Economic Regulation." *Journal of Law and Economics* 19:211-240.

Ray, E.J. 1981a. "The Determinants of Tariff and Nontariff Restrictions in the United States." *Journal of Political Economy* 89:105-121.

_____. 1981b. "Tariff and Nontariff Barriers to Trade in the United States and Abroad." *Review of Economics and Statistics* 63:161-168.

_____. 1990. "Protection of Manufactures in the United States," in David Greenaway, ed., *Global Protectionism: Is the U.S. Playing on a Level Field*. New York, NY: MacMillan.

Saunders, R. S. 1980. "The Political Economy of Effective Tariff Protection in Canada's Manufacturing Sector." *Canadian Journal of Econonomics* 13:340-348.

Tollison, R. D. 1982. "Rent Seeking: A Survey." *Kyklos* 35:575-602.

U.S. Department of Commerce, Bureau of the Census. Various years. *Annual Survey of Manufactures*. Washington, D.C.

U.S. Department of Commerce, Bureau of the Census, 1987. *Census of Manufactures, 1989: Concentration Ratios in Manufacturing*, 1991. Washington, D.C.

U.S. International Trade Commission. 1990a. *Estimated Tariff Equivalents of U.S. Quotas on Agricultural Imports and Analysis of Competitive Conditions in U.S. and Foreign Markets for Sugar, Meat, Peanuts, Cotton, and Dairy Products*. USITC Publication 2276. Washington, D.C.

U.S. International Trade Commission. 1990b. *The Economic Effects of Significant U.S. Import Restraints, Phase II: Agricultural Products and Natural Resources*. USITC Publication 2314. Washington, D.C.

6

The Competitiveness of U.S. Food Exports in the Japanese Market

Thomas W. Hertel

Introduction

The Japanese market is of critical importance to a broad range of U.S. food exporters. In 1988 this market absorbed roughly 12 percent of global food imports, and approximately 21 percent of U.S. food exports. Thus, a disproportionate share of U.S. food exports go to Japan. Furthermore, due to growing economic and political pressures on the Japanese food sector, this market is destined to expand significantly in the future (ABARE 1988). The purpose of this chapter is to examine the responsiveness of an index of U.S. competitiveness in the Japanese food market to exogenous changes in: (a) current import barriers, and (b) unit costs of food and nonfood production in the U.S., Japan, and globally, under alternative assumptions about market structure.

The term "competitiveness" has experienced a great rise in popularity over the last decade. Indeed, the IATRC devoted an entire symposium to this theme (see Bredahl, Abbott, and Reed 1994). There remains considerable debate about what "competitiveness" is and how it should be measured. However, one thing seems clear. This is a *relative* concept which is only meaningful if one defines a basis for comparison. Thus, when addressing the issue of the competitiveness of U.S. food exports, one must first answer the question: Competitive relative to what? Is the appropriate comparison across different goods within the same region (i.e. the comparative advantage concept), or is it across similar goods supplied from different regions (the absolute advantage concept)?

Comparative advantage is appropriate for determining which goods will be exported and which will be imported over the long run. It will also play a role in some of the general equilibrium simulations presented here. However, the main focus of this chapter will be on changes in a sector-specific competitiveness measure for U.S. food exports to Japan.

Perhaps the most obvious sector-specific measure involves a comparison of relative prices. Hayes *et al.* (1991) discuss this approach for the case of beef trade. Their findings hinge crucially on the point in the marketing chain at which prices are measured. When farmgate prices are used, South America, Australia, and New Zealand are more competitive than the U.S., but this is not necessarily so when prices for wholesale cuts are considered. They advocate the use of prices for wholesale cuts, adjusted for transportation costs.

There are several problems with this type of simple price comparison. First of all, as Hayes *et al.* (1991) point out, manufactured food products tend to be differentiated. Thus, in comparing U.S. and Australian beef exports, one is not necessarily comparing homogeneous goods. Secondly, if product differentiation is endogenous, and motivated by heterogeneous preferences on the part of consumers, then one way to erode a competitor's market share is to leave prices unchanged but to increase varieties. In other words, simple price comparisons fail to capture the effect of changes in the number of varieties in a given market, despite the fact that this is an important dimension of competitiveness.

This chapter proposes an alternative measure of competitiveness which takes its cue from the consumer. After all, it is ultimately the consumer who determines which products will be purchased. It will be argued that the logical basis for determining the competitiveness of U.S. exports to Japan is not a simple price comparison, but rather to compare the price of U.S. exports in that market *relative to consumers' unit expenditure*. The latter measures the cost of consumers attaining a certain level of utility, given the current offering of products and prices in the market. This unit expenditure function is an explicit function of consumer heterogeneity, inherent biases for Japanese products, and the number of varieties available, *as well as relative prices*.

The next section of this chapter develops this competitiveness index in some detail. Not surprisingly, market share is shown to be an increasing function of the index. Perhaps less obvious is the result that this index is also a key piece of information for imperfectly competitive firms producing differentiated food products. In particular, optimal markups are shown to be an increasing function of a firm's own competitiveness index. In sum, this index is a central measure of market conditions facing U.S. food exporters.

This chapter then presents some qualitative, partial equilibrium analysis of the impact which tariff cuts for farm and food imports into Japan are likely to have on the competitiveness of U.S. food exports. The implications of technological progress in the global food sector are also examined. Empirical estimates of these effects are then developed using an applied general equilibrium model of global trade. This has the advantage of endogenizing the response of U.S. competitors in the Japanese market, as well as permitting an analysis of the impact of nonfood technological progress on the competitiveness of U.S. food exports to Japan. The chapter concludes with a summary of major findings.

Household Preferences, Unit Expenditure and the Competitiveness Index

Let us begin by formally defining $C(i,r,s)$ as the *competitiveness index* for producers of i, from region r, in the sth market:

$$C(i,r,s) = PD(i,s)/PDS(i,r,s). \tag{1}$$

It is the ratio of unit expenditure on i in region s, PD(i,s), to the price to consumers of the particular good i supplied by producers in region r, PDS(i,r,s). For example, if we are interested in examining the competitiveness of U.S. food exporters in the Japanese market, then i = food, r = USA, and s = Japan. In this case, when C ($\cdot$) increases, U.S. food exporters become relatively more competitive in the Japanese market, thereby increasing their market share. Note that this competitiveness index is only useful when products of type i are *differentiated*. If they were instead homogeneous, then $C(i,r,s) = 1 \ \forall$ r,s and there is no scope for changes in its value.

What kind of differentiated-products unit expenditure function makes sense for PD(i,s)? This is dictated by the nature of consumer preferences. (For an extensive treatment of alternative models of product differentiation see Beath and Katsoulacas 1991.) The appropriate specification surely varies across different types of products. In this chapter, emphasis is placed on assessing competitiveness in processed food products. Thus it is useful to consider a situation in which product differentiation arises from heterogeneous tastes for the characteristics embodied in different varieties of food products. Following Anderson, DePalma, and Thisse (1989), consider the following utility function for an individual consumer consuming Q_v of v = 1 ....V different varieties of food:

$$U_v(Z;Q_v) = \ln Q_v + Y - P_v Q_v - \gamma \sum_{k=1}^{m} (Z^k - Z_v^k)^2; \qquad v = 1...V \tag{2}$$

Here, the variables Z^k describe the m characteristics of the consumer's ideal variety. This ideal variety differs from the actual characteristics of variety v, as described by Z_v^k. The sum of the squares of these differences, weighted by γ, gives the utility "penalty" associated with consumption of variety v, as opposed to the ideal variety. Once the consumer has determined the optimal variety, he/she chooses the quantity to be purchased in order to maximize utility. (Expenditure (Y) is invariant to price.) This is given by $Q_v^*=P_v^{-1}$.

If consumers' ideal characteristics are distributed according to the multinomial logit, then Anderson *et al.* (1989) show that the consequences for aggregate behavior can be summarized in the form of a CES utility function with elasticity of substitution $\sigma=\rho/(1-\rho)=\mu^{-1}$, where the parameter μ is the standard deviation of consumer tastes. Thus, as consumer tastes become more diverse (μ increases), $\rho\rightarrow 0$ and the aggregate utility function collapses to the Cobb Douglas form. Conversely, for homogenous preferences ($\mu\rightarrow 0$), $\rho\rightarrow 1$, and $\sigma\rightarrow\infty$. This aggregate behavior is conveniently summarized in the following CES unit expenditure function:

$$PD = [\sum_{v\in VARIETIES} PDS(v)^{(1-\sigma)}]^{\left(\frac{1}{1-\sigma}\right)} \tag{3}$$

Unit expenditure, PD, is increasing in any individual price, PDS, but decreasing in the total number of varieties available. That is, at constant prices and expenditure, the representative consumer is able to attain a higher level of utility if more varieties are available.

In order to relate the unit expenditure function above to data on bilateral trade of commodity i among *regions*, we must make the assumption that all firms in any given region, r, charge the same price, and, furthermore, that these firms are active in any region s, where sales from r appear. In this way, we obtain a revised version of (3) in which prices are indexed over regions, and each price is weighted by the number of firms operating in each region. This yields:

$$PD(i,s) = [\sum_{r\in REG} N(i,r)(\alpha(i,r,s)*PDS(i,r,s))^{(1-\sigma)}]^{\left(\frac{1}{1-\sigma}\right)} \tag{4}$$

The "shift" parameter $\alpha(i,r,s)$ has been introduced in order to facilitate calibration of the model to observed regional variation in expenditure shares. It reflects factors not included in the model (i.e., considerations other than transport costs and policy wedges -- see also Venables 1987). Among other things, it captures the possibility that Japanese consumers may prefer Japanese products, *ceteris paribus*. Note, however, that the elasticity of substitution among varieties is *invariant* to their origin.

Having established a particular form for PD(i,s), we are now in a position to explore the role of the competitiveness index (1) in determining market share. Partial differentiation of (4) with respect to the price of food varieties supplied from PDS(i,r,s) gives rise to a market share equation:

$$[QDS(i,r,s)/QD(i,s)] = N(i,r)*\alpha(i,r,s)^{(1-\sigma)}*[PD(i,s)/PDS(i,r,s)]^{-\sigma} \quad (5)$$

where QD(i,s) is the aggregate demand for commodity i in s, and QDS(i,r,s) is the share supplied by producers in r. Note that this share is *increasing* in the competitiveness index, as expected.

Proportionate differentiation of (5) gives:

$$qds(i,r,s) = qd(i,s) + \sigma*c(i,r,s) + n(i,r) \quad (6)$$

where the lower case is used to denote proportionate change in upper case variables, i.e. pd=dPD/PD and c(i,r,s) is the proportionate change in the competitiveness index as defined in (1). Equation (6) shows sales to s by producers in r increase proportionately with the overall size of the market, qd(i,s). It also increases with the introduction of additional firms/varieties into the market, *conditional on* c(i,r,s) = 0. Finally, increases in competitiveness translate, through σ, into larger market share. Thus, for products where consumer preferences are more homogeneous, such that $\sigma \to \infty$, a small change in competitiveness can have a big impact on the market share of firms based in r, in market s.

In order to better appreciate the components of the competitiveness index, it is useful to substitute in the proportionate change form of PD(i,s) to obtain:

$$c(i,r,s) = \left[\theta(i,r,s) - 1\right] pds(i,r,s) + \sum_{k \neq r} \theta(i,k,s)\, pds(i,k,s)$$

$$- \left[\sigma-1\right]^{-1} \sum_{k} \theta(i,k,s)\, n(i,k) \quad (7)$$

Here, $\theta(i,r,s)$ represents the expenditure share for product i by consumers in s devoted to goods from region r. Because unit expenditure itself depends on pds(i,r,s), a 1 percent decline in this variable generates a less than 1 percent increase in c(i,r,s). Similarly, unit expenditure is itself decreasing in n(i,r), so that a 1 percent increase in n(i,r), *ceteris paribus*, will lead to a less than proportionate increase in market share.

Before leaving this discussion of the unit expenditure function and the competitiveness index, it is useful to consider the special case whereby *products are differentiated by origin rather than by firm*. This is a variant of the so-called Armington approach to trade modeling, which is generally used in conjunction with the assumption of perfect competition. In this case the number of firms plays no role in (4). It may simply be absorbed into the constant, $\alpha(i,r,s)$. Thus, $n(\cdot)$ drops out of (6) and (7), and the competitiveness index collapses to a simple ratio of the sum over *all* share-weighted price changes (unit expenditures) relative to the price of the *single* differentiated product supplied from r. In this way, the perfect competition/Armington model may be viewed as a special case of the model developed in this chapter.

Firm Behavior, Market Structure, and Competitiveness

I must next confront the question: How do firms differentiate their products? This chapter follows in the tradition of Dixit and Stiglitz (1977). In particular, I postulate that manufacturers of food products make fixed, recurrent outlays on advertising and R&D in order to establish a new variety in the marketplace. Once this is accomplished, production of the product itself is subject to constant returns, so that marginal cost equals average variable cost. However, average total cost (fixed outlays included) declines with increasing sales. Firms cover the costs of product differentiation -- perhaps even earning excess profits -- by marking up price over marginal costs:

$$\text{MKUP}(i,r) = \text{PS}(i,r)/\text{MC}(i,r), \tag{8}$$

where MKUP, PS, and MC denote the markup, producer price, and marginal cost of production respectively for variety i from region r. (Note that I am assuming *integrated markets* such that markups cannot be tailored to individual markets.) In differential form, recognizing that marginal cost equals average variable cost, I have:

$$\text{ps}(i,r) = \text{avc}(i,r) + \text{mkup}(i,r). \tag{9}$$

Price Linkages

In order to move from ps(i,r) to pds(i,r,s), it is necessary to introduce several price linkage equations. These are summarized in Figure 6.1. At the top of that figure are the ex-factory *agents' supply prices*, ps(i,r), which

FIGURE 6.1 Price Linkages in the Model (Proportionate Changes)

$ps(i,r)$ $= avc(i,r) + mkup(i,r)$: **optimal markup**

$\downarrow$

$pfob(i,r,s) = ps(i,r) - ts(i,r,s)$: **destination taxes**

$\downarrow$

$pcif(i,r,s) = \gamma(i,r,s)\ pfob(i,r,s) + [1 - \gamma(i,r,s)]pt$: **international transport**

$\downarrow$

$pms(i,r,s) = pcif(i,r,s) + tm(i,r,s)$: **tariffs**

$\downarrow$

$pds(i,r,s) = pms(i,r,s) + td(i,r,s)$: **commodity taxes**

---

are determined by the optimal markup equation (9). The combination of output and export taxes results in a set of *ad valorem destination taxes*, $ts(i,r,s)$, which account for the difference between the price producers receive and *fob* prices. The next price linkage condition establishes *cif* prices by destination on the basis of route-specific transportation margins, $[1-\gamma(i,r,s)]/\gamma(i,r,s)$. The last two price linkage relationships in Figure 6.1 incorporate the effects of tariffs, $tm(i,r,s)$ $(r\neq s)$, and domestic commodity taxes, $td(i,r,s)$.

Implications of a Change in Protection

Now consider the impact of a uniform reduction in the Japanese rate of protection for food imports, such that $tm(i,k,s) = tm$, $\forall\ k \neq s$, $i = $ food, and $s = $ Japan. How does this affect $c(i,r,s)$? In order to keep things simple, it is convenient to make some partial equilibrium assumptions. (These will be relaxed in the empirical section below.) In particular, assume that input and international transport costs are not affected (avc $= pt = 0$). Also, assume that the *ad valorem* form of other policies are unchanged, so that $ts(i,r,s) = tds(i,r,s) = 0$. Then we may substitute the price linkages in Figure 6.1 into (7) to obtain:

$$
\begin{aligned}
c(i,r,s)\,|_{avc=pt=0} = &\left[\theta(i,r,s) - 1\right]\left[tm + \gamma(i,r,s)\,mkup(i,r)\right] \\
&+ \sum_{k \neq r,s} \theta(i,k,s)\left[tm + \gamma(i,k,s)\,mkup(i,k)\right] \\
&+ \theta(i,s,s)\,mkup(i,s) \\
&- \left[\sigma-1\right]^{-1}\sum_{k} \theta(i,k,s)\,n(i,k)
\end{aligned}
\tag{10}
$$

But how is tm likely to influence $mkup(i,r)$? This depends on the form of the optimal markup expression. If, for example, foreign firms did not adjust their markups [$mkup(i,r) = 0 \; \forall \; r \neq s$], and domestic firms decided to reduce theirs at the same rate as the tariff [$mkup(i,s) = tm$], *then it is possible that the competitiveness index would not change* (provided $n(i,r) = 0$). For this reason it is important to examine the behavior of markups.

Optimal Markups

The optimal markup over marginal cost for producers of i in r is determined by their total perceived elasticity of demand as follows:

$$
MKUP(i,r) = ET(i,r)/[ET(i,r)-1].
\tag{11}
$$

This may be totally differentiated to obtain:

$$
mkup(i,r) = [1-MKUP(i,r)]et(i,r).
\tag{12}
$$

Since $MKUP(i,r) \geq 1$, this implies a negative relationship between $et(i,r)$ and $mkup(i,r)$. That is, an increase in the perceived demand elasticity will cause the optimal markup to fall.

Under the integrated markets hypothesis, this perceived demand elasticity (ET) is a quantity-weighted average of the perceived demand elasticities, $E(i,r,s)$, in each of the individual market destinations:

$$
ET(i,r) = \sum_{s \in dest} SSHR(i,r,s)E(i,r,s),
\tag{13}
$$

where $SSHR(i,r,s)$ is the share of sales of i from r to market s. Proportionate differentiation of (17) gives the following:

$$
et(i,r) = \sum_{s \in dest} ELSHR(i,r,s)\{[qs(i,r,s)-qo(i,r)] + e(i,r,s)\}.
\tag{14}
$$

Here, $ELSHR(i,r,s) = SSHR(i,r,s)E(i,r,s)/ET(i,r)$ represents the contribution of market s to the (i,r) firm's total perceived demand elasticity. Also, $dSSHR(i,r,s)/SSHR(i,r,s) = [qs(i,r,s) - qo(i,r)]$, which is the difference between the proportional change in sales to market s and total sales. Recognizing that

$$\sum_{s \in DEST} ELSHR(i,r,s) = 1,$$

we can remove qo(i,r) from the summation to obtain:

$$et(i,r) = -qo(i,r) + \sum_{s \in DEST} ELSHR(i,r,s) [qs(i,r,s) + e(i,r,s)]. \qquad (15)$$

Thus, even if $e(i,r,s) = 0 \; \forall \; s$, ET(i,r) may change if the policy shock causes sales to be shifted between markets with differing perceived demand elasticities.

At this point the prospects for a simple form of the competitiveness index, even in partial equilibrium [i.e. (10)], look rather bleak. However, note that if the share of total sales to Japan from region r is small ($SSHR(i,r,s) \approx 0$) then $ELSHR(i,r,s) \approx 0$. This is an empirical question to which Table 6.1 provides an answer for aggregate manufactured food sales. From this table it is clear that the Japanese market is only "large" for Japanese producers. Thus we expect that the primary effect of the tariff shock will be to influence domestic firms' markups. Rewriting (10) with the restriction that mkup $(i,k) = 0 \; \forall k \neq s$ yields:

$$c(i,r,s)\big|_{avc = pt = mkup (i,k \neq s) = 0} = \theta(i,s,s)[mkup(i,s) - tm]$$
$$- [\sigma - 1]^{-1} \sum_{k} \theta(i,k,s) n(i,k). \qquad (16)$$

TABLE 6.1 Share of Total Manufactured Food Sales to the Japanese Market

Region of Origin (r)	SSHR (food, r, Japan)
Australia	.011
North America	.004
Japan	.857
Korea	.058
EEC	.004
ROW	.008

Source: Salter-II data set (Jomini *et al.* 1991) as modified by Hertel (1993).

Perceived Demand Elasticities

In order to understand how Japanese domestic markups might change in response to a tariff, we must be more specific about the nature of inter-firm rivalry. For purposes of this chapter, I limit my attention to static, noncooperative behavior. In this context there are two standard avenues available. In the first, firms conduct their "thought experiment" about changes in QDS(i,r,s) in response to a perturbation in PDS(i,r,s) based on the (Bertrand) assumption that rival firms will leave their price unaltered. This results in the following perceived demand elasticity (for detailed derivations see Hertel 1994):

$$E(i,r,s) = \sigma - (\sigma-1)[\Theta(i,r,s)/N(i,r)]. \tag{17a}$$

The alternative approach is to assume that firms take their rivals' quantities as given, assuming instead that rivals will adjust their prices to clear the markets for differentiated products. This (Cournot) assumption results in a smaller perceived demand elasticity, the formula for which follows:

$$E(i,r,s) = \sigma/\{1+(\sigma-1)[\Theta(i,r,s)/N(i,r)]\}. \tag{17b}$$

In order to understand how $E(i,r,s)$ changes as a function of changes in market shares and numbers of firms, we totally differentiate (17) to obtain the following expression:

$$e(i,r,s) = \{[1 - \sigma]\Theta(i,r,s) / D(i,r,s)\}shr(i,r,s) \tag{18}$$

where the form of D ($\cdot$) depends on the nature of inter-firm rivalry. The Bertrand and Cournot forms of D($\cdot$) are given by:

$$D_B(i,r,s) = N(i,r)E(i,r,s) \tag{19a}$$

and

$$D_C(i,r,s) = N(i,r) + \Theta(i,r,s)[\sigma(i) - 1] \tag{19b}$$

The variable shr(i,r,s) is the proportionate change in a representative firm's (value-based) market share $\Theta(i,r,s)/N(i,r)$. Since $\Theta(i,r,s) = N(i,r)*[\alpha(i,r,s)*PD(i,s)/PDS(i,r,s)]^{(\sigma-1)}$ we have:

$$shr(i,r,s) = [\sigma-1][pd(i,s) - pds(i,r,s)]. \tag{20}$$

Substituting (20) into (18), we obtain the following expression for the individual firm's perceived demand elasticity in market s:

$$e(i,r,s) = -(1-\sigma)^2[\Theta(i,r,s)/D(i,r,s)]\,c(i,r,s). \tag{21}$$

Thus the perceived demand elasticity *itself* depends on the competitiveness index. In particular, a decline in competitiveness, $c(i,r,s) < 0$, raises $e(i,r,s)$ and lowers mkup(i,r). In the context of (16) this means that we expect some of the decline in tariff to be offset by a decline in Japanese producers' domestic markups. However, since mkup(i,s) is also a function of $n(i,s)$, it is conceivable that markups could *rise* due to the *exit* of domestic firms, thereby *reinforcing* the tariff cuts direct effect on U.S. firms competitiveness. For this reason we must now turn to the firms' entry/exit decision.

Firm Entry/Exit

Thus far, nothing has been said about the determination of $n(i,k)$, the number of firms active in each of the regions. In some cases, barriers to entry/exit may cause this to be fixed, (i.e., $n(i,k) = 0\ \forall k$) in which case $c(i,r,s)$ is essentially determined by [mkup(i,s) - tm]. Hertel (1994) discusses this no entry case in some detail and shows that, independent of the nature of the static, noncooperative game played by domestic firms, *markups will fall*. Furthermore, they will fall by more, the more imperfectly competitive the domestic market conditions. Finally, the decline in the domestic markups will never dominate the tariff cut, and the competitiveness of U.S. food manufacturers will rise, but not by as much as would be the case in the absence of this "procompetitive" effect of the tariff cut on domestic markups.

While the no entry/exit cases may be of interest in some instances, particularly when the tariff cut is relatively modest, a significant cut in the tariff on food imported into Japan would likely drive some domestic firms out of business. To capture this effect, we may introduce the following zero profit condition:

$$ps(i,s) = atc(i,s) = scatc(i,s) - \Omega(i,s)\,qf(i,s). \tag{22}$$

This states that firms must cover *all* costs including the fixed, recurrent costs associated with differentiating their product. The proportional change in average total cost may be broken into two components: the change in scale-constant average total cost (scatc) and that associated with changes in total output (qf). By increasing output/sales firms can spread their R & D/marketing costs over more units and lower atc.

In partial equilibrium, scatc = 0 by virtue of exogenous input costs. Thus, the only avenue open to Japanese food producers in a longer run, zero profits equilibrium, is to rationalize by increasing $qf(i,s)$. With total demand falling:

$$qo(i,s) = qf(i,s) + n(i,s) < 0,$$ (23)

this requires $n(i,s) < qo(i,s) < 0$. On the other hand, the proportional change in output required of foreign firms will be minimal, since $SSHR(i,r,s) \approx 0$ so that $n(i,r) \approx 0 \;\; \forall r{\neq}s$.

It can be shown that in this type of partial equilibrium with entry, the impact of a tariff cut on the domestic markup (and hence price, since avc = 0), depends entirely upon the uncompensated elasticity of demand (ε) for the composite commodity (Hertel 1994). In most of the theoretical work in this area, aggregate preferences are assumed to be of the Cobb Douglas form, so that $\varepsilon = -1$. In this case the change in domestic variety required to satisfy (21):

$$n(i,s) = (\sigma-1)\{[1 - \theta(i,s,s)] / \theta(i,s,s)\} \, tm,$$ (24)

precisely offsets the effect of lower foreign prices, leaving unit expenditure unchanged (Hertel 1994). This results in stationary domestic markups and unchanging domestic prices (i.e., $mkup(i,s) = ps(i,s) = 0$). This is merely a reflection of the fact that *perturbations to this partial equilibrium system with entry and unitary demand elasticity are observationally equivalent to a perfectly competitive industry* (Hertel 1994, Proposition Two). This enables us to readily deduce the change in $c(i,r,s)$ for $r{\neq}s$ as follows:

$$c(i,r,s)_{avc\,=\,pt\,=\,mkup\,(i,k{\neq}s)\,=\,0}^{entry/\,exit} \equiv pd(i,s) - pds(i,r,s)$$ (25)

$$= -pds(i,r,s) = -tm.$$

Implications of a Change in Technology

Next, consider the impact of an improvement in food manufacturing technology which lowers costs. Once again, I will assume that the cost of international transport is not affected (pt = 0). Furthermore, now *all* policies are assumed unchanged ($ts(i,r,s)=tds(i,r,s)=tm(i,r,s)=0$). Substitution of the price linkage relationships in Figure 6.1 into (7) yields the following expression for the change in competitiveness of producers of i form r in region s:

$$c(i,r,s) = [\theta(i,r,s) - 1] \gamma(i,r,s) [avc(i,r) + mkup(i,r)]$$
$$+ \theta(i,s,s) [mkup(i,s) + avc(i,s)]$$
$$+ \sum_{k \neq r,s} \theta(i,k,s) \gamma(i,k,s) [avc(i,k) + mkup(i,k)] \qquad (26)$$
$$- (\sigma - 1)^{-1} \sum_{k} \theta(i,k,s) n(i,k)$$

From (26) it may be seen that the change in competitiveness of producers depends critically on their cost reduction *relative to* those in the export market (s) and in competitor regions (k≠r,s). Consider first the case where costs fall in *all* regions by avc(i,r)= avc < 0. Because optimal markups, like demand, are homogenous of degree zero in prices, this kind of shock will leave markups unchanged. Assuming mkup = pt = 0, as well as no entry [n(i,r) = 0], and setting γ(i,r,s) = γ $\forall$r≠s (for simplicity), (25) collapses to the following:

$$c(i,r,s)_{mkup = pt = 0}^{no\ entry} = (1 - \gamma) \theta(i,s,s) avc \qquad (27)$$

In this case global technical changes *erodes* the competitiveness of all exporters, relative to domestic producers. This is because the price cuts are not fully transmitted into the export market, owing to the international transport component of the *cif* price. Finally, if the cost reduction is specific to region r, then competitiveness will be enhanced, since only the first term on the right hand side of (26) will be altered.

What if the technological change applies only to the international transport sector? In this case avc = 0, but pt < 0. Now the cost reduction works to the advantage of foreign producers, with the size of the gain in competitiveness depending on the relative importance of international transport in the *cif* price of its product. The larger [1 - γ(i,r,s)] the greater the gain in competitiveness. If γ(i,r,s) = γ $\forall$r≠s then:

$$c(i,r,s)_{mkup = avc = 0}^{noentry} = (\gamma - 1) \theta(i,s,s) pt \qquad (28)$$

which is the complement of (26). Therefore, if pt = avc, then c(i,r,s) = 0.

Empirical Model

In order to assess the empirical significance of changes in either technology or Japanese trade policy, we must turn to a data base with bilateral trade flows and bilateral, commodity-specific international transportation margins. For purposes of this study, I employ the SALTER-II data base (Jomini *et al.* 1991). Hertel (1993) has further modified this data base to incorporate variety-specific fixed costs and optimal markups, consistent with the model of firm-specific product differentiation developed above.

The data base is *global* in the sense that the addition of a "Rest of World" (ROW) region assures exhaustive coverage of global production and trade. This is important because U.S. producers compete not only with Japanese food manufacturers, but also with other exporters in the Japanese market. This implementation of the SALTER-II data base aggregates regions into six groups: Australasia (Australia & New Zealand), U.S.-CN (United States and Canada), Japan, Korea, EEC-12, and ROW. (For ease of exposition, I will equate the changes in the U.S.-CN region's competitiveness to changes in U.S. competitiveness.)

The entries in Table 6.2 refer to key behavioral parameters in the model for the nine sector aggregation used in this study. The first column reports elasticities of substitution in production (i.e. in value-added), by sector. The next column reports the assumed degree of substitutability in consumption by product category. The values of σ reported in Table 6.2 vary by sector, ranging from a low of 3.8 for the service sectors to a high of 10 in the case of transportation equipment. These technology and preference parameters are assumed to be the same for all regions. However, individual regions' supply and demand behavior differs due to differing expenditure shares.

The final two columns in Table 6.2 refer to the calibrated, initial markups in the imperfectly competitive sectors. Since both agriculture and natural resources are treated as perfectly competitive, this entry is not applicable. Based on econometric evidence reported by Hall (1988), it is postulated that manufacturing and services producers markup price above marginal cost. The third column in Table 6.2 reports the Lerner indices, (P-MC)/P, for the U.S. These estimates are based on econometric results reported in Domowitz, Hubbard, and Peterson (1988). The service sectors are treated as monopolistically competitive.

Markups for other regions are obtained by deriving the relationship between optimal markups in the U.S. and the share of fixed costs in value-added under a zero profits equilibrium. This share is assumed to be constant across regions, so that regional variation in markups reflects underlying differences in the share of value-added in total costs. The combination of markups and σ's in Table 6.2 rules out Bertrand behavior for most sectors. Therefore, Cournot behavior is assumed in this model. The implied numbers of firms in the U.S. and Japan are given in the final

TABLE 6.2 Parameters Used in the Model

Sector	Elasticities of Substitution*		Lerner Indices		Cournot-Equivalent Numbers of Firms	
	Value-Added	Consump-tion	U.S.	Japan	U.S.	Japan
Agriculture	0.56	4.4	n.a.	n.a.	n.a.	n.a.
Natural Resources	1.12	5.6	n.a.	n.a.	n.a.	n.a.
Food	1.12	4.4	.29**	.25	11	36
Textiles, Clothing and Footwear	1.26	7.0	.29**	.23	4	8
Non-ferrous Primary Metals	1.26	5.6	.28**	.26	6	7
Transport Equipment	1.26	10.0	.26**	.21	3	3
Other Manuf	1.26	5.6	.34**	.31	4	5
Construction	1.40	3.8	.27	.27	100	100
Services	1.26	3.8	.27	.27	100	100

Notes: * Source: SALTER parameter file (Jomini *et al*. 1991).
       ** Source: Domowitz *et al*. (1988) as modified by Hertel (1993).

columns of Table 6.2. Because the calibrated Japanese food manufacturing is very close to the monopolistically competitive lower bound, as dictated by σ, the Cournot-equivalent number of firms is very large in this sector. Since the results reported are potentially sensitive to this value, the implications of using smaller values of N in the Japanese food sector will also be explored.

General Equilibrium Results

Tariff Cuts

The first type of shock which I will consider involves a reduction in the tariff on Japanese imports. Simply on the basis of back-of-the-envelope calculations, we expect that a 1 percent reduction in the Japanese tariff on food will lower Japanese market prices for U.S. food exports by about one percent. Since the foreign share of the market for processed food is small this will translate into almost a one percent reduction in the cost of U.S. food relative to all Japanese food--in the absence of changes in the number of varieties available. Thus, U.S. food exports to Japan should increase by about $\sigma * 1$ percent = 4.4 percent.

Using this *naive* prediction as a benchmark, turn to the first column in Table 6.3 which reports the results of a one percent tariff cut when the number of firms in each regional food sector is fixed [n(i,r) = 0 $\forall$i,r]. Note first of all that the competitiveness index for U.S. exports in the Japanese market rises by 0.85 percent, which is less than one percent. This is due to the fact that Japanese unit expenditure for food *falls* by - 0.13 percent. [The numeraire in this experiment is the Japanese consumer price index.] About half of this decline is due to a (-0.07 percent) decline in the price of domestic Japanese food products in response to lower demand. [Note that the change in the Japanese markup is negligible due to the large number of Cournot-equivalent firms (N=36).] The remaining part of this decline in unit expenditure may be attributed to the decline in the price of imports. As a result, U.S. sales to this market increase by less than 4.4 percent (3.15 percent for a 1 percent tariff cut).

When zero profits are enforced via entry/exit of firms/varieties, this presents an additional source of change in Japanese consumers' unit expenditure. The variable v(food, Japan) in the fifth row of Table 6.3 reports the percentage change in the total number of food varieties on offer in the Japanese market. The decline of -0.21 percent reported in the fourth column of Table 6.3 reflects the dominant effect of domestic firms exiting the market. Because unit expenditure is *decreasing* in variety, this exit forces consumers to spend more, *on average*, in order to reach the same level of utility at unchanged prices. From (7) it may be seen that $-(\sigma-1)^{-1}$ is the multiplicative factor on v($\cdot$) so that the difference in unit expenditures owning to the varietal change is $-(\sigma-1)^{-1} (-0.21) = 0.06$, which accounts for the difference between pd(food, JP) in the entry and/no entry cases. Because pd falls less, the relative competitiveness of U.S. producers increases more in the case where domestic firms exit. In other words, since U.S. producers now have fewer varieties with which to compete, they can increase their combined market share still further (3.44 percent vs. 3.15 percent in the no-entry case).

For purposes of comparison, the last column in Table 6.3 shows the impact of a 1 percent cut in the rate of protection on Japanese *agricultural* imports. Since these serve as an important input into the food processing sector, domestic firms are able to lower their prices, thereby *eroding* the competitiveness of U.S. exporters in the market [c(food, U.S.,JP) = -0.15 percent]. [Because this tariff reform lowers the price of the numeraire (the Japanese *cpi*), the *relative* price of U.S. exports rises (0.03 percent).] Consequently, U.S. producers lose market share and U.S. food sales to Japan fall. In light of the fact that most liberalization proposals involve reductions in the tariffs on both raw and processed products, it is interesting to compare the last two columns in Table 6.3. It appears that, *on average*, provided the cut in raw product tariffs is not more than 6 times the cut in processed product tariffs, the competitiveness of U.S. food exports will increase. Of course, the pattern of protection varies widely across disaggregate commodities, and such an *average* figure will not be helpful in predicting the effects of piecemeal reforms. For this purpose further commodity disaggregation is required.

The second and third columns in Table 6.3 provide an insight into the potential range of *procompetitive effects* as a result of lowering food product tariffs in Japan. In these two cases, the *initial equilibrium* Cournot-equivalent number of firms is reduced -- but held constant during the course of the simulation (i.e. *no entry*). (It is unnecessary to conduct such sensitivity analysis for the entry/exit case, since we have already noted that the procompetitive effect of tariff cuts in that case will be offset by declining variety, thereby leaving markups little changed.) This is an important sensitivity analysis, since (12)-(19) show that the change in Japanese optimal markups depends on N and evidence from the Japanese food industry suggests a substantially greater degree of concentration (Sutton 1991). This comparison of alternative *no entry* results in Table 6.3 shows that indeed reductions in N *do* cause markups to fall more, thereby diminishing the competitive advantage of U.S. producers as tariffs are reduced. However, the change is not too great. This is because of the relatively small value of σ. If this parameter is increased the procompetitive effect is rapidly strengthened [see equation (21)].

TABLE 6.3 Implications of a 1 Percent Reduction in Japanese Tariffs

Variable	No Entry/Exit			Entry/Exit	Entry/Exit
	N = 36	N = 15	N = 3	N = 36	N = 36
c(food,U.S., JP)	.85	.84	.74	.93	-.15
mkup (food,JP)	-.00	-.01	-.08	.00	.00
pd*(food,JP)	-.13	-.14	-.24	-.06	-.12
pd*(food,U.S.,JP)	-.98	-.98	-.98	-.99	.03
v(food,JP)	0	0	0	-.21	-.01
qs(food,U.S.,JP)	3.15	3.10	2.79	3.44	-.45
qo(food,U.S.)	.03	.03	.03	.03	-.01

Note: * Price changes are percentage changes *relative to* the Japanese consumer price index.

Technological Change

Now, consider the effects of Hicks-neutral technological change on the competitiveness of U.S. food exports in the Japanese market. Numerical results from these simulations are presented in Table 6.4. The first pair of columns in this table corresponds to the case whereby the *global* food sector experiences a 1 percent rate of technological progress. Based on the *no entry* partial equilibrium analysis reported above, we expect this to erode the competitiveness of food imports, with a larger decline the greater is the international transport margin. This is reflected in the no entry result reported in the first column of Table 6.4, whereby there is a -0.14 percent decline in competitiveness for a 1 percent rate of change in Hicks-neutral technical change, globally.

When entry is permitted, unit expenditure falls more rapidly (-1.35 percent versus -1.18 percent under no entry). This is because unit expenditure is *decreasing* in variety. While this diminishes the competitiveness of any *given* U.S. food manufacturer, it need not result in a decline in U.S. sales to Japan. This is because the number of U.S firms increases (+0.47 percent). Indeed, the decline in qs(food, U.S., JP) is only one-third larger than in the no-entry case, despite a doubling of the decline in any given firm's competitiveness.

The next pair of columns in Table 6.4 report the results of a 1 percent rate of technical change which occurs *only in the U.S. food sector*. This one percent reduction in costs translates into a one percent (1.01 percent to be exact) increase in the competitiveness of U.S. producers in the Japanese market, and a 1.20 percent increase in U.S. food output. When additional U.S. producers are permitted to enter the market, Japanese food manufacturers are allowed to exit, U.S. sales increase *directly* due to the relatively larger number of active firms, and *indirectly* due to the decline in aggregate variety which tends to raise Japanese unit expenditure. Thus qs(food, U.S, JP) increases by 5.45 percent as opposed to 4.22 percent under no entry/exit.

The final pair of columns in Table 6.4 relate to an experiment which was not discussed above, namely a shock to technology in the nonfood (and non-agricultural) sectors of the Japanese economy. This is interesting from a *general equilibrium* point of view, since it tends to draw resources away from the Japanese food sector, *thereby diminishing its competitiveness*. It is also an interesting experiment to consider since technical change in Japanese nonfood manufacturing has tended to outstrip that in the food sector during the postwar period. Turning first to the no entry case, we see that a 1 percent rate of growth in Japan's nonfood sector lowers the *cpi* in that region, relative to other regions. Thus, the *relative* price of U.S. food exports rises (0.41 percent). Because Japan's food sector must compete with the more efficient nonfood sector, its costs rise. Thus unit expenditure on food rises even more than pds(food, U.S., JP), thereby enhancing the competitiveness of U.S. food exporters. This raises U.S. market share. In addition, the economic growth engendered by widespread technical change in Japan translates into increased demand for manufactured food products. As a result, U.S. exports rise by 1.79 percent.

When entry/exit is permitted, a 1 percent rate of nonfood technological progress in Japan leads to the exit of Japanese food processors, and the entry of food processors in other regions. This serves to further enhance U.S. competitiveness. Sales to Japan now rise by 2.73 percent for every 1 percent increase in Japanese nonfood technical change. Comparing this figure to that in the second pair of columns, we conclude that a 2 percent rate of technical change in the Japanese *nonfood* sector generates about the same growth in U.S. exports to Japan as does a 1 percent rate of technical change on the part of U.S. *food* manufacturers.

TABLE 6.4 Implications of a 1 Percent Rate of Technological Change in Various Sectors Under Entry/Exit

	Source of Technological Change					
	Global Food Production		U.S. Food Production		Japanese Nonfood Production	
Variable	No Entry	*Entry*	No Entry	*Entry*	No Entry	*Entry*
c(food, U.S., JP)	-0.14	-0.30	1.01	1.07	0.11	0.36
pd(food, JP)	-1.18	-1.35	-0.04	-0.02	0.52	0.65
pds(food, U.S., JP)	-1.04	-1.05	-1.05	-1.09	0.41	0.29
n(food, JP)	0	0.49	0	-0.09	0	-0.31
n(food, U.S.)	0	0.47	0	1.02	0	0.07
v(food, JP)	0	0.47	0	-0.07	0	-0.29
qs(food, U.S., JP)	-0.29	-0.37	4.22	5.45	1.79	2.73
qo(food, U.S.)	0.75	0.77	1.20	1.25	0.04	0.04

Note: Entries are percentage change in selected endogenous variables.

Summary and Conclusions

This chapter has proposed the use of a new index for assessing changes in the competitiveness of differentiated, processed food product exports in a specific market. The numerator of this index consists of consumers' unit expenditure on the composite of all goods of a given type, in a given market (eg., Japan). This is divided by the price paid by consumers in that market for imports from a given region (eg., the United States). The strengths of this proposed competitiveness index are several. First, it is a consumer-based measure which takes its cue from the structure of consumer preferences for a given type of food product. Furthermore, it recognizes that most food products are not homogeneous. In this chapter preferences are treated as heterogeneous, thereby motivating the demand for differentiated food products. Another strength is that the index takes into account the role of changes in the number of varieties of products on offer in the marketplace. In particular, endogenous product differentiation implies that the competitiveness of U.S. food exports in the Japanese market can change, even if the price of U.S. products relative to Japanese products is unchanged.

Partial equilibrium analysis of changes in this competitiveness index reveals a number of important points. First of all, in a no-entry equilibrium, an across-the-board reduction in Japanese tariffs on food products will have a less-than-proportionate impact on the competitiveness of U.S. exports. This is due to the resulting decline in aggregate unit expenditure, as well as the associated procompetitive effect of the tariff cut on domestic Japanese markups. By contrast, in the presence of a zero-profits, entry/exit equilibrium, there is a tendency for the effect of declining domestically-supplied varieties to offset the impact of lower import prices, leaving unit expenditure, as well as domestic markups, unchanged. Thus tariff reductions and competitiveness tend to move in lock-step.

Changes in food and nonfood technology also have an important impact on the competitiveness of U.S. exports in the Japanese market. The direction of this effect depends on the sectoral and spatial distribution of the technical progress. While region-specific improvements in U.S. food manufacturing technology improve competitiveness, a global improvement in food manufacturing efficiency actually *erodes* U.S. competitiveness in the Japanese market, due to the incomplete price transmission caused by international transport costs. By contrast, if the technical progress occurs in the nonfood sectors, the opposite conclusions may be reached. For example, an increase in nonfood productivity, relative to food sector productivity in the Japanese economy, improves the competitiveness of U.S. food manufacturers in that market.

This chapter uses an applied general equilibrium model of global trade to illustrate these points. Key features include the use of bilateral trade and international transport margins data, as well as a comprehensive treatment of imperfect competition in the U.S., Japan, and other regions. Global coverage means that behavior of U.S. competitors is also explicitly considered, along with competition between food and nonfood sectors for region-specific endowments. Numerical results confirm the qualitative insights discussed above. A one percent cut in Japanese processed food tariffs results in a 0.93 percent increase in U.S competitiveness under entry/exit, and a 0.74 percent increase in competitiveness under the no entry/exit closure. The technical change simulations reveal that a two percent rate of technical progress in *Japanese nonfood production* has about the same effect on U.S. exports to Japan as does a one percent rate of technical change in *U.S. food manufacturing*. This highlights the value of a general equilibrium approach to the analysis of competitiveness and U.S. food exports to Japan.

Notes

1. This equation is derived by assuming a competitive transport sector with the composite cost index PT.
2. The results in Hertel (1994) assume domestic firms do not export their products. This fits well with the case of Japanese food manufacturing.
3. We have reason to expect that the absolute value of the uncompensated price elasticity of demand for most food products in industrialized economies is less than one, i.e. $0 > \varepsilon > -1$. What happens in this case? Now the effect of the decline in domestic varieties will *dominate* the effect of lower foreign prices and domestic prices will actually *rise*. This causes a *strengthening* of the competitive position of U.S. exporters.
4. About 40 percent of all U.S. processed food sales to Japan are intermediate inputs used by the Japanese food sector.

References

ABARE. 1988. "Japanese Agricultural Policies: A Time of Change," Policy Monograph No. 3, Australian Bureau of Agricultural and Resource Economics, Canberra.

Anderson, S.P, A. DePalma, and J. Thisse. 1989. "Demand for Differentiated Products, Discrete Choice Models, and the Characteristics Approach." *Review Economics Studies* 56:21-35.

Beath J. and Y.S. Katsoulacas. 1991. *The Economic Theory of Product Differentiation*. Cambridge: Cambridge University Press.

Bredahl, M., P.C. Abbott, and M.Reed. 1994. *Competitiveness in International Food Markets*. Boulder, CO: Westview Press.

Dixit, A.K. and J.E. Stiglitz. 1977. "Monopolistic Competition and Optimum Product Diversity." *American Economic Review* 67:297-308.

Domowitz, I, R.G. Hubbard, and B.C. Petersen. 1988. "Market Structure and Cyclical Fluctuations in U.S. Manufacturing." *Review of Economics and Statistics* 70:55-66.

Hall, R.E. 1988. "The Relation Between Price and Marginal Cost in U.S. Industry." *Journal of Political Economy* 96:921-947.

Hayes, D.J., J.R. Green, H.H. Jensen, and A. Erbach. 1991. "Measuring International Competitiveness in the Beef Sector," *Agribusiness: An International Journal* 7:357-374

Hertel, T.W. 1993. "Introducing Imperfect Competition into the SALTER Model of Global Trade." Department of Agricultural Economics Staff Paper #93-4. West Lafayette, IN: Purdue University.

_____. 1994. "The Procompetitive Effects of Trade Liberalization in a Small, Open Economy". *Journal of International Economics* 36:391-411.

Jomini, P., J.F. Zeitsch, R. McDougall, A. Welsh, S. Brown, J. Hambley, and J. Kelly. 1991. "SALTER: A General Equilibrium Model of the World Economy." Canberra, Australia: Australian Industry Commission.

Sutton, J. 1991. *Sunk Costs and Market Structure*. Cambridge, MA: MIT Press.

Venables, A.J. 1987. "Trade and Trade Policy with Differentiated Products: A Chamberlinian-Ricardian Model." *The Economic Journal* 97:700-717.

7

The Impact of CUSTA on Canada's Tomato Processing Industry: Tariffs, Technical Regulations, and Industry Bargaining Behavior

Erna van Duren and Clarissa de Paz

Introduction

Soon after the Canada-U.S. Trade Agreement (CUSTA) was signed in 1987 many observers of the Canadian tomato processing industry predicted the latter's imminent decline. The combination of changes in tariffs and technical regulations resulting from CUSTA, in conjunction with the annual tomato processing contract negotiations, conducted by the Ontario Vegetable Growers Marketing Board (OVGMB) and Ontario tomato processors (under the umbrella of the Ontario Food Processors Association (OFPA)), were deemed to be obstacles that the industry could not survive. It is six years later and the industry continues its struggle for survival given the changes in tariffs, technical regulations, and contract bargaining behavior that have occurred since 1987. This chapter assesses the impact of these three sets of policy variables on the competitiveness of the Canadian tomato processing industry using a multi-region, multi-product spatial equilibrium model of the North American industry.[1] Competitiveness is defined here as the "sustained ability to profitably gain and/or maintain market share" (Task Force on the Competitiveness of Canada's Agrifood Industry 1990).

To set the stage for our analysis, the next section provides a brief "tour" of the tomato processing industry in North America. The subsequent section summarizes key changes that have occurred in tariffs, technical regulations, and industry bargaining behavior in the Canadian industry since CUSTA. We then briefly describe a spatial equilibrium model and provide analysis using that model. The final section offers concluding comments.

The Tomato Processing Industry: A Brief "Tour"

North America accounts for well over half of the world's production of tomato processinges, which has surpassed 20 million metric tons in recent years. California alone accounts for over 45 percent, while the U.S. midwest, Canada (Ontario) and Mexico account for approximately 4 percent, 3 percent and 2 percent, respectively. The European Community (EC) accounts for approximately 35 percent of world production.

Although fresh and processed tomatoes are somewhat interchangeable in final consumption, the processing and fresh tomato industries operate separately since their genetic material and processing-distribution channels are quite different. The variety of processed tomato products available in North America continues to burgeon. The basic categories of processed products are paste, sauce, juice and whole-peel.

Tomato paste is the most fungible of all processed tomato products; and given California's dominance in producing tomatoes for paste (about 75-80 percent of all tomato processinges), the California paste price largely drives processed product and raw tomato pricing in North America. Stocks and production in California are particularly important to the paste price. Expectations about production and stock behavior in the rest of the world play a secondary influence.

The North American tomato processing industry comprises three regions: the U.S. west of the Mississippi (mainly California), the U.S. east of the Mississippi, and Canada.[2] In the U.S. west, production is centered in the Central Valley area of California, while in the U.S. east it is clustered in the mid-western U.S. states of Ohio, Indiana and Illinois. In Canada, production is concentrated in southwestern Ontario, around Leamington (over 95 percent of tomatoes grown for processing in Canada). Both Canada and the U.S. import tomato products, while the U.S. is also a significant exporter of some products.

The tomato processing industry in the U.S. west is mature and spans all vertical components. Breeding of genetic material (transplants and seeds), growing raw tomatoes, primary processing, and further processing or "remanufacturing" are all well developed industries.

The industry in the U.S. midwest has evolved into a remanufacturing center which ships tomato paste in from California and processes it into consumer-ready forms such as sauces, drinks, and paste to serve the significant population centers in the eastern U.S. The U.S. midwest also draws on its own raw tomato supply to supplement the more cost effective paste from California. It is more self-reliant for whole-peel tomatoes, which are considerably more expensive to transport.

In the past, the Canadian tomato processing industry has included many of the vertical components of the more fully developed California industry. Raw tomatoes have been grown in Ontario and processed into all types of tomato products for the Canadian market. Paste was imported from the U.S. and other countries to complement paste from local tomato production, and as well imports of whole-peel products from non-U.S. sources have also been signficant. Until recently the genetic material was imported from the southeastern U.S. in the form of tomato transplants.

Policy and Industry Behavior Changes Induced by CUSTA

CUSTA has induced changes in three sets of policy/behavioral variables that are important to the functioning and competitiveness of the Canadian tomato processing industry: tariffs, technical regulations and industry bargaining behavior through the interaction of the OVGMB and the tomato processors under the OFPA.

Tariffs

Before CUSTA, Canada and the U.S. protected their tomato industries from foreign competition through similarly structured tariffs of approximately the same level. CUSTA phases out this protection, which ranges from 7.5 percent to 15 percent on bilateral trade over 10 years.

Technical Regulations

In the food industry, technical regulation refers to rules dealing with food safety, quality and consumer information that must be followed by producers, processors and sellers as they procure inputs, develop, process, distribute and market products (van Duren 1992).

Over 40 percent of the technical regulations that affect Canada's tomato processing industry have marketing related objectives such as grading, packing, labelling, transportation, promotion and distribution. Consumer protection in the market place, health and safety in the workplace and the administrative requirements involved in regulating are objectives of 20 percent of the regulations. Consumer information and consumer health are objectives of 15 percent of the regulations, while only 5 percent are concerned with bilingualism, and less than 3 percent with environmental protection.

Nearly 55 percent of the regulations that affect the tomato processing industry are directed at the agrifood industry in general, while just over 40 percent are directed at horticulture. Approximately 10 percent are aimed at the beverage industry, while just under 3 percent are intended for all economic activity. The tomato processing industry is affected by regulations that are targeted at various vertical levels of the agrifood, horticulture and beverage industries.

Processors are the target of approximately 75 percent of the regulations that affect the industry, while firms involved in wholesale and distribution are the target of just over 30 percent. Input suppliers to firms at all levels of the vertical chain are the target of approximately 10 percent, as are firms outside the horticulture industry. Retailers are the target of approximately 8 percent of the regulations, while farmers and hotels, restaurants, and institutions are targeted by less than 5 percent of the regulations.

Consumers are not targeted by any of the regulations that affect the tomato processing industry, but they are affected by over 20 percent of them. Whether an industry level was affected by a technical regulation was determined by assessing whether it would affect the conduct, cost, revenue, or willingness to conduct that activity. Levels of the industry that are not targeted by technical regulation can be affected by spillovers resulting from their content and application at related vertical levels, and also in industries that are horizontally related.[3] Processors are affected by over 95 percent of the regulations, followed by firms involved in distribution and wholesaling, which are affected by just over 40 percent, and input suppliers, which are affected by just under 25 percent of them. Retailers, firms outside the industry, farmers and firms at the HRI level of the industry are affected by 14 percent, 12 percent, 8 percent, and 5 percent of technical regulations, respectively.

Institutional Behavior

Tomato contracts between processors and growers are negotiated under different institutional arrangements in the three regions of the North American tomato processing industry.

In the U.S. midwest, processors negotiate contracts with growers individually. Pricing, delivery, discounts, and premiums, as well as product characteristics, are addressed in these contracts. Desirable raw product characteristics differ for tomatoes destined for paste versus whole-peel products.

In California, prices for tomato processinges are negotiated between processors and the California Tomato Growers' Association using a "value equating" price approach.[4] The "equating" price is based on the soluble solids of tomatoes, which is determined using third party grading. Thus, tomatoes with a higher level of soluble solids (i.e. 5.5 percent) receive a higher price than those with lower levels (i.e. 5 percent). Although soluble solids are not the only desirable characteristic for tomato processinges they are by far the most important for paste, the California industry's main product (van Duren *et al.* 1989). The contracts typically contain a base price, a minimum and/or maximum price, deduction standards, rejection standards, time premiums and/or penalties, and quality premiums and/or penalties.

In Ontario, tomato contracts are negotiated under the legislative framework established by the Ontario Farm Products Marketing Act. This Act requires the annual establishment of a Negotiating Agency which must contain members of the OVGMB and tomato processors. The Negotiation Agency conducts negotiations about prices and various terms relating to product characteristics, conditions, and charges relating to production and marketing of tomato processinges, among others. If an agreement is not negotiated by February 28 each year, unresolved issues are sent to an Arbitration Board, which is established under the auspices of the Ontario Farm Products Marketing Commission, for resolution through final offer arbitration. Composition of the Negotiation Agency and the terms which it can negotiate have proven to be quite flexible. Therefore, the negotiation process and the terms of the agreement between processors and growers in the Ontario tomato processing industry have changed substantially since CUSTA was signed, thereby allowing the industry to become more competitive with its U.S. counterpart (Martin and van Duren 1990; van Duren and McFaul 1993).

A Spatial Equilibrium Model of
the North American Tomato Processing Industry

In order to assess the impact of changes in tariffs, technical regulations, and industry bargaining behavior in Ontario on industry level indicators of competitiveness, a spatial equilibrium model of the North American tomato processing industry was constructed.

The model is based on three regions and four tomato products. Processed tomato products which are consumed at the retail level are paste, sauces, juices and drinks, and whole-peel products. At the farm level, raw tomatoes for processing into paste, sauces, juices and drinks, and whole-peel products are considered. Tomatoes for consumption as fresh tomatoes are not included because they are produced with different technologies and generally by different farmers. Products in the model are represented by the subscript j. At the farm level, the subscript t is appended to indicate raw tomatoes.

The U.S. west of the Mississippi forms the first region in the model. Production prices are for San Francisco, while consumption prices are for Los Angeles. The U.S. east of the Mississippi is the second region, in which production prices are for Fremont, Ohio and consumption prices are for New York. Canada forms the third region in the model. Production prices are for Leamington, while consumption prices are for Toronto. Regions in the model are represented by the superscript i (k is the anchor region -- the U.S. west).

All quantities in the model are measured in tomato paste equivalents (using an industry standard of 31 percent solid levels). Since the tomatoes grown in the three regions of North America are used for a different mix of end-uses under different climatic and soil conditions, we used different raw product to paste equivalent yield factors for each of the products in each of the three regions. In Canada, these factors are based on raw tomatoes having a solid content of 5 percent, while in the U.S. east and U.S. west, solids contents of 4.8 percent and 5.5 percent are used, respectively.

Retail Demand

Consumption for processed tomato products is treated as a function of an intercept and the price of processed tomato products and fresh tomatoes. Cross-price effects were considered. Data on the per capita consumption of these products for 1987-89 and price indices for 1987-89 in the U.S. and Canada, as well as cross-sectional price data for Ontario in 1991, were used to generate starting values for the retail demand equations in the three regions. Then information from industry sources and various publications was used to improve the coefficients in this block of the model.[5] Consumption is given as:

$$D_j^i = a_j^i + b_j^i \, P_j^i$$
$$\text{where } a_j^i > 0$$
$$b_j^i \lessgtr 0 \tag{1}$$
$$P_j^i \geq 0$$

Subscripts j indicate products (paste, sauce, juice/drinks, whole-peel, fresh), superscripts i indicate regions (Canada, U.S. east and U.S. west [indicated by k]), subscript t indicates raw tomatoes for processing, and all products are measured in tomato paste equivalents.

Farm Supply

Production of raw tomatoes for processing is treated as a function of an intercept and the price received for tomatoes for processing into four different end-uses (paste, sauce, juices/drinks, and whole-peel). Cross-price effects are not considered. Data on acreage and yields for all tomato processinges for 1980-89 were used to generate econometric estimates that did not account for the end-use of the tomatoes. Information from industry sources and various publications was used to allocate production to the four end-uses for each region. As well, it was used to further improve the coefficients in this block of the model. The supply function is:

$$S_{jt}^i = c_{jt}^i + d_{jt}^i \, P_{jt}^i$$
$$\text{where } c_j^i \gtreqless 0$$
$$d_{jt}^i > 0 \qquad\qquad\qquad\qquad (2)$$
$$P_{jt}^i \geq 0$$

Transfer Costs

The model contains two types of transfer costs: transportation and policy induced transfer costs, most notably tariffs. Therefore, the transfer cost block of the model is in two parts. Again, information from industry sources and various publications was used to compile as much of the transfer costs block as possible. Missing elements were estimated by calculating the relevant distances among the various production and consumption points in the three regions, calculating adjustment factors and applying them to the known elements. All tariff information was readily available. Since interregional trade in raw tomatoes does not occur, due to the product's perishability, transfer costs for raw tomatoes were not included in the model. However, since transport of raw tomatoes to the processing plant within a region is part of the net price that processors pay, it is included in the processing cost block of the model. The transfer cost policies are:

$$TC_j^{ik} = P_j^k - P_j^i$$ (3)

$$\text{where } TC_j^{ik} \geq 0$$

$$P_j^i = P_j^k + TC_j^{ik}; \ P_j^k < P_j^i$$

Processing Costs

Processing costs in this model contain several components since they reflect the difference between the price paid for raw products and the price received from consumers. This approach was necessary in this first iteration of construction of the model because data on this sensitive piece of information are virtually impossible to obtain and because the availability of other data only allowed the generation of farm level supply and retail demand equations. Fortunately, discussion with a food industry consultant who specializes in brokerage and business planning, discussion with industry sources and an intensive review of any relevant literature allowed the generation of processing costs for each product for each region. The model contains adjustment factors for processing costs, so that as additional information becomes available this block of the model can be improved easily. Processing costs are:

$$PC_{jt}^i = P_j^i - P_{jt}^i$$

$$\text{where } P_{jt}^i > 0$$ (4)

Storage Demand

Storage of tomato products plays a critical role in the tomato processing industry, especially paste, which is stored aseptically in tanks. Unfortunately, as the industry continues to consolidate and rationalize in Canada and the U.S. the availability of storage data continues to decrease. However, storage demand has been incorporated in the model. In the current version of the model, storage demand is based on an intercept and the expected price of end-use tomato products. Expected prices (indicated by an *) have been set equal to those in the base solution to the model, since we did not conduct any multi-year policy impact simulations. Storage demand is given by:

$$SD_j^i = e_j^i + f_j^i P_j^{i*}$$ (5)

Other

The model of the North American tomato processing industry also contains several other variables such as trade, several economic surplus measures, and competitiveness indicators. These are discussed in the next section to the extent required to set the stage for the results of our analysis of the impacts of policy changes induced by CUSTA on the competitiveness of Canada's tomato processing industry.

Solving the model requires substituting the relevant processing and transfer costs into the retail demand and farm level supply functions:

$$D_j^i = a_j^i + b_j^i (P_j^k + TC_j^{ik}) \tag{6}$$

$$S_{jt}^i = c_{jt}^i + d_{jt}^i (P_j^k + TC_j^{ik} - PC_{jt}^i) \tag{7}$$

By applying the simple equilibrium condition that retail demand plus storage demand minus supply across all regions is equal to zero in any one year, the model can be expressed as a system of linear equations. Solving this system of equations produces a vector of retail prices in the U.S. west, which can then be used to solve for the remainder of the model.

The equilibrium condition is:

$$D + SD - S = 0 \tag{8}$$

which is equivalent to:

$$(a_j^i + e_j^i - c_{jt}^1) + (b_j^i - d_{jt}^i)P_j^k + (b_j^i - d_{jt}^i)TC_j^{ik} + d_{jt}^i PC_{jt}^i + f_j^i P_j^{i*} = 0,$$

which can be written in matrix notation for n=4 products and m=3 regions:

$$\beta\, P^k = \alpha$$

where $\beta = \begin{bmatrix} b_j^i - d_{jt}^i & \cdot\cdot \\ \cdot & \cdot \\ \cdot & \cdot \\ \cdot & b_m^n - d_{mt}^n \end{bmatrix}$ $\tag{9}$

and,

$$
\alpha = \begin{bmatrix}
(a_j^i + e_j^i - c_{jt}^i) + (b_j^i - d_{jt}^i)TC_j^{ik} + d_{jt}^i PC_{jt}^i + f_j^i p_j^{i*} \cdots \\
\cdot \qquad \cdot \\
\cdot \qquad \cdot \\
\cdot \qquad \cdot \\
(a_m^n + e_m^n - c_{mt}^n) + (b_m^n - d_{mt}^n)TC_j^{nk} + d_{mt}^n PC_m^n + f_m^n P_m^{n*}
\end{bmatrix}
$$

Base Solution

Constructing a synthetic spatial-equilibrium model with data as limited as they are for the tomato processing industry makes model validation difficult. Table 7.1 contains a comparison of the production and consumption results in the base solution to similar actual data for 1987-89. These are the only data available for validating the model. The first four columns indicate the percent of total North American production and consumption accounted for by each region in the model for the four end-use products, and the percentage points by which the base solution is over or under the actual. With the exception of juice consumption in all regions and whole-peel consumption in the U.S. midwest, the base solution is within 3 percentage points of the actual, and generally much closer. Total production quantities in the base solution as a percent of the actual data range from 98 percent to 100 percent, while the consumption estimates are not quite as good, ranging from 103 percent to 107 percent of the actual data.

Table 7.2 contains the results of the base solution. It contains several variables which can be used to assess the impacts of policy changes induced in Canada on the competitiveness of the Canadian tomato processing industry. Variables relating to profitability and market share form the core of the set of variables used for this analysis because they reflect the definition of competitiveness which is used by most participants in Canada's agrifood industry: "The sustained ability to profitably gain and maintain market share."

Prices: Retail and farm level prices in each region for each product are indexed to 1 in the U.S. west for each industry level. These estimates are consistent with the pattern of prices that is observed for these tomato products.

Production and Consumption: Production and consumption in each region are reported as a percent of the North American total for each product, and total quantities as well in thousands of metric tons (MTs). These results have been discussed in reference to Table 7.1.

TABLE 7.1 Comparison of Actual Production and Consumption Data to the Base Solution

	Percent of North American Total				Quantity (000MTs)
Production	Paste	Sauce	Juice	Whole Peel	Total
CANADA					
-Actual	3	5	1	3	43
-Base Over/Under	-1	0	2	-1	100%
Actual (+/-)					
U.S. EAST					
-Actual	7	13	2	7	107
-Base Over/Under	-1	- 3	-1	2	98%
Actual (+/-)					
U.S. WEST					
-Actual	90	82	96	90	1179
-Base Over/Under	3	3	-1	-1	99%
Actual (+/-)					
U.S.					
-Actual	97	95	99	97	1286
-Base Over/Under	1	0	-2	1	99%
Actual (+/-)					
Consumption					
CANADA					
-Actual	6	6	24	11	78
-Base Over/Under	-1	0	-8	-3	103%
Actual (+/-)					
U.S. EAST					
-Actual	65	65	51	56	760
-Base Over/Under	1	0	18	-3	103%
Actual (+/-)					
U.S. WEST					
-Actual	29	29	25	33	351
-Base Over/Under	0	0	-10	6	107%
Actual (+/-)					
U.S.					
-Actual	94%	94	76	89	1111
-Base Over/Under	1	0	8	3	104%
Actual (+/-)					

TABLE 7.2 Base Solution Summary

		Percent of North American Total				Quantity (000MTs)
		Paste	Sauce	Juice/ Drinks	Whole- Peel	Total
Retail Prices[*]	Canada	1.3	2.4	4.6	4	NA
	U.S. East	1.2	2.2	4.3	3.7	NA
	U.S. West	1	1.7	3.1	2.7	NA
	N.A.	NA	NA	NA	NA	NA
Farm Prices[*]	Canada	1.6	2.8	5.2	4.5	NA
	U.S. East	1.3	2.3	4.4	3.8	NA
	U.S. West	1	1.6	2.8	2.3	NA
	N.A.	NA	NA	NA	NA	NA
Production [**]	Canada	2	4	3	2	43
	U.S. East	6	10	1	9	105
	U.S. West	92	86	96	89	1163
	N.A.	98	96	97	98	1268
Consumption [**]	Canada	5	6	16	9	80
	U.S. East	65	65	69	52	780
	U.S. West	30	29	15	39	377
	N.A.	95	94	84	91	1157
Trade/Production Ratio (in percent)	Canada	-107	-22	-333	-198	-78
	U.S. East	-836	-570	-5472	-420	-621
	U.S. West	59	63	85	60	62
	N.A.	7	-2	21	17	6
Consumers Surplus[***]	Canada	5	4	36	11	$ 398
	U.S. East	83	86	61	60	$3603
	U.S. West	12	10	3	29	$3655
	N.A.	95	96	95	89	$4093
Farmers Surplus[***]	Canada	3	8	5	5	$ 254
	U.S. East	7	13	1	11	$ 438
	U.S. West	90	79	94	84	$3655
	N.A.	97	92	95	95	$4093
Value Added to Farm Gate Cost of Raw Products (in percent)	Canada	3	7	15	13	9
	U.S. East	31	44	380	50	49
	U.S. West	24	30	30	42	33
	N.A.	25	32	36	43	35
Processors' Revenues[***]	Canada	$ 32	$ 273	$ 43	$ 117	$ 466
	U.S. East	$ 119	$ 668	$ 51	$ 457	$1295
	U.S. West	$1371	$3600	$ 755	$2728	$8454
	N.A.	$1490	$4268	$ 806	$3185	$9749
Processors' Costs[***]	Canada	$ 31	$ 256	$ 37	$ 104	$ 428
	U.S. East	$ 91	$ 463	$ 11	$ 304	$ 869
	U.S. West	$1103	$2760	$ 580	$1922	$6365
	N.A.	$1194	$3223	$ 591	$2226	$7234

Note:
[*] Indices, with U.S. West = 1 for paste.
[**] Percent of North America for products, total in 1,000 metric tons.
[***] Based on indices, so $ value is not truly relevant, percent of North America for products.

Market Share Measures: Market share, as indicated by the regional share of production variable discussed above, and the trade production ratio (net exports/production) are the two indicators of market share that can be used to assess the impact of policy changes due to CUSTA on the industry's competitiveness. The trade/production ratio is for North American trade; it does not account for trade with third countries. Given its construction, the signs in the base solution are correct, as are the relevant magnitudes (again with the exception of juice). Since the net trade/production ratio reflects the impact on trade and production it must be interpreted carefully; one needs to know the direction of trade in the base solution to do so correctly. A net importer would see the trade production ratio increase if net imports increased, while a net exporter would see the ratio increase if net exports increased.

Consumers Surplus: The economic welfare of consumers is presented by providing a total value (based on price indices, which means it does not have a relevant dollar value) and the percentage of total consumers' surplus for each product that accrues to consumers in each region. We use consumers' surplus in our analysis because of the purported benefits of policy changes induced by CUSTA to Canadian consumers (lower prices, more product choice, better value).

Farmer Surplus: The economic welfare of the farmers who grow tomatoes for processing is presented by providing a total value (based on price indices, which means it does not have a relevant dollar value) and the percentage of total farmers' surplus for each product that accrues to producers in each region. The cost of raw product to the processor, which is equal to farmers' revenue, is a less satisfactory indicator of farm level profitability. However, it can provide additional insight into to the impact of policy changes induced by CUSTA on Canadian farmers, and, therefore, we include it.

Value Added Above Raw Product Cost: Processors' performance is assessed by calculating the total value added above total raw product cost as a percent of total raw product cost. The stream of revenues and costs considered includes domestic retail sales, costs, and revenues realized from storage (these are not discounted because we only use the model for one year of analysis), exports, and raw product costs.

Processors' revenues and their *raw product costs* are reported since information on the components of value added can assist in understanding the impact of policy changes induced by CUSTA on Canadian tomato processors.

Analysis and Results

We conducted several simulations to assess the impacts of policy changes induced by CUSTA on Canada's tomato processing industry. Changes in tariffs, technical regulations, and industry bargaining behavior were considered. A brief description of how the simulations were conducted within the model and the key results are presented in Table 7.3. Results for the U.S. east and U.S. west are not provided in the table, but are referred to in the discussion.

Tariffs

The tariff reductions required under CUSTA were simulated by setting all tariffs on products entering Canada equal to zero. Since the tomato industry produces several products, the maximum and minimum effects on prices at the retail and farm level are reported. They decline to 90.8 percent and 89.4 percent of the value in the base solution, for both levels, respectively. Sauce prices decrease most, while whole-peel prices decrease least. The retail and farm price effects are the same because no change in other processing costs was considered in this simulation.

Canada's share of the North American market decreases by 0.3 percent, while the trade-production ratio increases by 50.5 percent, reflecting an increase in net imports.

Farmer surplus decreases significantly, to 83.6 percent of its level in the base solution, while consumers' surplus increases to 122.9 percent of the base. Processors' value added does not change because retail and farm level prices are affected similarly by the tariff reductions, and because effects are not large.

Technical Regulations

The effects of three types of technical regulations which could be reduced through CUSTA were simulated: food grades and standards, labelling, and import certificates. The simulation for import certificates also serves as a proxy for other regulations which offer protection at the border, such as inspection.

i) Food Grades and Standards: The impacts of Canadian food grades and standards are generally too small to be important. The only impact of enforcing them on domestic products but not on imports would be a minor improvement in consumers' surplus (to 102 percent of the base).

ii) Labelling: The impacts of Canadian labelling regulations are very small, but sufficient to consider how they could be improved. The effect that they have through higher processing costs is negligible (+/- 0.1

percent), but by constraining consumers from purchasing products with attributes they deem to be important, they have a negative effect on consumer surplus (94.2 percent of the base). To the extent that foreign products do not enter Canada because they do not meet the requirements, net imports decline marginally (as indicated by the decline in the trade/production ratio by 6.7 percent).

iii) Import Certificates: The impact of Canadian import certificate (or other similar) regulations is the most significant of the technical regulations that we considered. Because these regulations operate at the border their effects are greater. Removing them would result in lower farm and retail prices. They would decline to approximately 96 percent of their base solution levels. Canada's market share would decline negligibly (by 0.1 percent), and net imports would increase (by 16.6 percent). Farmers' surplus would decline considerably, to 94.2 percent of the base solution level, while consumers' surplus would increase to 107.6 percent of the base. Processors' value added would not be affected because, again, the impact was passed through to farmers in the form of lower raw product prices.

iv) Total Impact: Considering the cumulative impact of the three technical regulations considered in this chapter provides some insight into their total impact on the competitiveness of Canada's tomato processing industry. However, since the effects of several other regulations were not simulated, these results should only be considered as partial.

As a group, the technical regulations we considered benefitted farmers and processors less than the tariff protection they enjoyed prior to CUSTA, and also had less impact on their competitiveness. Removing tariffs had less negative impacts on indicators of market share and profitability than removing technical regulations. However, removing tariffs also had a more beneficial impact on Canadian consumers than removing the technical regulations.

Industry Bargaining Behavior

In the simulations of the effects of tariffs and technical regulation, processors and farmers were assumed to be behaving as they had prior to CUSTA. This has not been the case, as was discussed earlier. To simulate the effect of changes in behavior at the farm and processing levels, and the manner in which the two levels have been approaching the goal of becoming competitive in the post-CUSTA business environment, we considered the following. First, through productivity pricing incentives, substantial increases in farm level productivity have

TABLE 7.3 Impact of Policy Changes Due to CUSTA on Prices and Industry Level Competitiveness Indicators in Canada

	Retail Prices Max Min	Farm Prices Max Min	Market Share	Trade/ Production*	Farmers' Surplus	Consumers' Surplus	Processors' Value Added
TARIFF REDUCTION UNDER CUSTA							
1. All tariffs into Canada equal to zero	90.8 89.4	90.8 89.4	-0.3	50.5	83.6	122.9	-0.1
TECHNICAL REGULATIONS							
Food Grade and Standards							
2. Increase in Processing Costs (0.5%)	100.0 100.0	100.0 100.0	-0.0	0.0	100.0	100.0	-0.0
3. Decrease in Processing Costs (0.5%)	100.0 100.0	100.0 100.0	0.0	-0.0	100.0	100.0	0.0
4. Increase in Retail Demand (0.5%)	100.1 100.0	100.1 100.0	-0.0	2.3	100.1	102.0	-0.0
5. Increase in Processing Costs (0.5%) & increase in Retail Demand (0.5%)	100.1 100.0	100.0 100.0	-0.0	2.3	100.0	102.0	-0.1
6. Decrease in Processing Costs (0.5%) & increase in Retail Demand (0.5%)	100.1 100.0	100.1 100.0	0.0	2.2	100.1	102.0	-0.0
Labelling							
7. Increase in Processing Costs (2.0%)	100.0 100.0	99.9 99.9	-0.0	0.2	99.8	100.0	-0.1
8. Decrease in Retail Demand (1.5%)	99.9 99.8	99.9 99.8	0.0	-6.8	99.8	94.2	0.1
9. Increase in Processing Costs (2.0%) & decrease in Retail Demand (1.5%)	99.9 99.8	99.8 99.7	-0.0	-6.7	99.6	94.2	0.0

Note: % of base solution except where indicated by *, which indicates change in %.

TABLE 7.3 (Continued)

	Retail Prices Max Min		Farm Prices Max Min		Market Share	Trade/ Production*	Farmers' Surplus	Consumers' Surplus	Processors' Value Added
Import Certificates									
10. Decrease in Transfer Costs into Canada (5.0%)	96.7	96.2	96.7	96.2	-0.1	16.6	94.2	107.6	-0.0
Total Impact									
11. Maximum (combination of 5,9,&10)	96.5	96.2	96.4	96.1	-0.1	12.1	0.0	103.5	-0.1
12. Minimum (combination of 6,9,&10)	96.5	96.2	96.6	96.2	-0.1	11.7	94.2	103.5	-0.1
Industry Bargaining Behavior									
13. Farm Level Productivity Incr. (15%)	100.0	100.0	100.0	100.0	0.1	-3.8	104.7	100.1	0.1
14. Farm Level Productivity Incr. (90%)	99.9	99.7	99.9	99.7	0.4	-20.7	130.4	100.5	0.4
15. Canadian Processing Margins Equal to those in the U.S. East	100.0	100.0	96.1	93.5	-0.1	8.1	91.3	100.0	-6.2
16. Canadian Processing Margins Equal to those in the U.S. West	100.0	100.0	85.3	77.2	-0.5	32.2	70.9	99.9	-25.9
17. Canadian Processing Margins Equal to those in the U.S. East and Farm Level Productivity Increase (15%)	100.0	100.0	96.1	93.4	-0.1	3.9	95.9	100.0	-6.2
18. Canadian Processing Margins Equal to those in the U.S. East and Farm Level Productivity Increase (90%)	90.8	89.4	90.8	89.4	-0.3	50.5	83.6	122.9	-0.1
19. Canadian Processing Margins Equal to those in the U.S. West and Farm Level Productivity Increase (15%)	100.0	100.0	85.3	77.2	-0.4	26.8	75.0	100.0	-25.8
20. Canadian Processing Margins Equal to those in the U.S. West and Farm Level Productivity Increase (90%)	99.9	99.8	85.1	77.0	-0.0	3.7	97.0	100.5	-25.4

Note: % of base solution except where indicated by *, which indicates change in %.

been realized. To simulate the effects of the average improvements to date and the nearly best improvement to date, we considered productivity improvements of 15 percent to 90 percent. In addition, since many processing firms operating in Canada are under pressure to be as profitable as their U.S. counterparts, we considered the effects of forcing the rest of the Canadian industry to respond to Canadian processors having processing margins equal to those for firms in the U.S. east and U.S. west.

The farm level productivity increases in Canada have to be very substantial before they affect North American prices, but they do create improvements in farmer surplus. Even at 15 percent it increases to 104.7 percent of its base solution. The decline in the trade-production ratio (by 3.8 percent) indicates net imports decrease, which is consistent with the tiny increase in market share (0.1 percent). Farm level productivity increases in Canada of 90 percent improve these indicators considerably, and even affect North American prices. Canada's market share improves by 0.4 percent and net imports decrease (as indicated by a decrease of 20.7 percent in the net trade/production ratio). Farmers' surplus increases significantly, to 130.5 percent of the base, as does processors' value added (by 0.4 percent).

Simulations of requiring Canadian processing margins to be equal to those in the U.S. east or west indicate that prices to farmers would have to decrease significantly if no price increases were passed on to consumers. Requiring margins in Canada to be equal to those in the U.S. east would result in farm prices declining to between 96.1 percent and 93.5 percent of their base, while requiring margins to be equal to those in the U.S. west would result in farm prices declining to 85.3 percent to 77.2 percent of their base. Such price decreases of this magnitude have already been observed for some categories of tomatoes in Canada (van Duren and McFaul 1993).

Simulation 20 indicates that farm-level productivity increases of more than 90 percent have to be achieved if the Canadian processing industry is to achieve processing margins comparable to those for the U.S. west without taking them out of farmers' profits and not lose North American market share.

Summary and Conclusions

This chapter has presented an industry level analysis of the impacts of three policy changes induced by the Canada-U.S. trade agreement on the competitiveness of Canada's tomato processing industry. The impacts of changes in tariffs, selected technical regulations, and industry bargaining behavior were assessed using a multi-region, multi-product, multi-level spatial equilibrium model.

First, tariffs offered significant protection to the Canadian tomato processing industry and their removal under CUSTA poses a significant competitive challenge for the Canadian industry.

Second, the cumulative impact of technical regulations on the competitiveness of the Canadian tomato processing industry could be substantial, although the effects of individual regulations are minor. Therefore, an assessment of the industry-level impacts of a more complete set of regulations should be conducted.

Third, the reduction in tariffs is likely more of a competitive challenge for the Canadian industry than potential changes in the technical regulations that may result under the on-going harmonization process being undertaken as a result of CUSTA. However, the industry-level impacts of a more complete set of technical regulations need to be simulated before this finding can be made conclusively.

Fourth, changes in the bargaining and other behavior of industry participants are critical in assessing the real impact of CUSTA on the industry's competitiveness. Improvements in industry relations, pricing mechanisms and productivity remain the key to survival and prosperity in the new North American, and global, business environment.

Notes

1. This chapter is the condensation of two longer papers: de Paz (1993) and van Duren and McFaul (1993). Both manuscripts are available from the Department of Agricultural Economics and Business, University of Guelph.

2. We have not yet considered Mexico as a separate region within the North American industry. It is treated as exogenous, like European and other countries.

3. For example, a regulation directed at butter may affect margarine.

4. The California Tomato Growers' Association is a voluntary organization. It is exempt from the Sherman Antitrust Act and is a qualified cooperative under the Capper Volstead Act.

5. A complete description of the model can be found in de Paz (1993).

References

Buxton, B.M and D. Roberts. 1992. "Economic Implications of Alternative Free Trade Agreements for the U.S. Fresh Tomato and Tomato Paste Industries." Selected Paper Presented at the American Agricultural Economics Association Annual Meeting, Baltimore, MD.

de Paz, C. 1993. "Assessing the Impact of Technical Regulation on the Ontario Tomato processing Industry." Unpublished M.S. thesis. University of Guelph.

George Morris Centre. 1993. "Principles and Actions on Standards and Regulation to Enhance the Competitiveness of Canada's Agri-Food Industry." Report presented to Agrifood Competitiveness Council.

Martin, L. and E. van Duren. 1990. " Reforming the Marketing System for Tomato processinges in Ontario." Prepared for the Industrial Restructuring Commission. Government of Ontario.

Sullivan, G.H. 1992. "Outlook for Tomato processing Production in North America." Paper presented to the Ministry of Agriculture and Food, Tomato processing Outlook at Ontario, Canada, January.

Sullivan, G.H. and F. Ravara. 1989. "Organization, Structure and Trade in the North American Tomato Processing Industry". Paper presented at the Third International Symposium on Tomato Processinges, Avignon, France.

Task Force of Competitiveness. 1990. "Report to the Ministers of Agriculture - Appendices; Working Group Reports." *Growing Together*.

van Duren, E. 1992. "The Impact of Technical Regulations in Canada's Agri-Food Industry: Results of a Pilot Study of Horticultural Processors." George Morris Centre Discussion Paper.

van Duren, E. and A. McFaul. 1993. "A Case Study of the Ontario Processing Tomato Industry's Response to the Canada-U.S. Trade Agreement." Department of Agricultural Economics and Business, University of Guelph.

van Duren, E., E. Dickson, D. Dupont and L. Martin. 1989. "The Impact of the Canada-U.S. Free Trade Agreement on Food Processing Sectors Supplied by Marketing Boards in Ontario." Prepared for the Industrial Restructuring Commissioner of Ontario.

8

Market Liberalization and Productivity Growth: An Empirical Analysis

Nicholas G. Kalaitzandonakes and Maury E. Bredahl

Introduction

Agricultural protection in most developed countries has distorted world markets and has had detrimental effects on the economic growth of agriculture in developing countries. Escalating budgetary costs coupled with growing friction between principal trading partners, such as the European Community, United States, and Japan, have generated pressures for comprehensive reform of agricultural policies. For the most part, however, agriculture has proven resistant to reform. Lack of political will to comprehensively liberalize agriculture has left protectionist policies intact in most developed countries. The reluctance of governments to liberalize agriculture can be attributed to fears that reform will lead to significant economic hardship and to eventual shrinkage of the protected agricultural industries.

Empirical evidence from agricultural industries that have been liberalized, however, reveals that economic reform need not necessarily cause shrinkage. In many instances reducing protection has led to increased production and industry growth (Koester 1991). Reducing protection and prices does not necessarily imply a downward movement along a stationary supply curve. Productivity gains resulting from liberalization may shift the supply curve sufficiently outward to expand rather than contract the reformed industries. These dynamic adjustments have been overlooked in traditional economic analyses of liberalization. Indeed, limited attention has been paid to the influence of protectionism (or liberalization) on productivity growth. Leibenstein (1966) has argued that protectionism tends to generate technical inefficiency and, thus, it is counter-productive. In contrast, Schultz (1979), Antle (1988), and Farrel and Runge (1983), emphasizing the influence of protectionism on technical change, have suggested that goverment protection leads to greater productivity growth as it lessens price uncertainty and provides incentives for innovation. Empirical evidence supporting any one of the above arguments has been quite limited (Van derMeer and Yamada 1990; Huffman and Evenson 1992; Fulginiti and Perrin 1993).

This study revisits the effects of protectionism and liberalization on agricultural productivity growth. First, a behavioral model explores the influence of liberalization on each of the three structural components of productivity growth -- technical change, technical efficiency, and scale efficiency. Second, an appropriate empirical model, where such influences can be explicitly shown, is discussed. Finally, a case study quantifies the effects of liberalization on productivity growth.

The Sources of Agricultural Productivity Growth

Any action that increases total output while holding all inputs constant is said to increase productivity. As discussed by Nishimizou and Page (1982), productivity growth can be decomposed into three separate components: (a) changes in technical efficiency, (b) changes in scale, and (c) technical progress. Figure 8.1 illustrates the decomposition. The production frontiers labeled F_1 and F_2 represent the "technically efficient" input-output combinations in two periods. Point A represents the initial position of a firm in period 1 and D its final position in period 2. Since the output per unit of input employed by the firm in period 2 increased relative to that in period 1, productivity has increased between the two periods. The illustrated growth in productivity can be decomposed into the three separate adjustments: First, improvement in technical efficiency is represented by the movement from point A to B. Second, an increase in the scale of production is depicted by the movement from point B to C. If this scale adjustment takes place over a production region where increasing (decreasing) returns to scale exist, productivity increases (decreases). Third, technological change results in an outward shift of the production frontier represented by a movement from point C to point D.

In agricultural production, technical change is most often embodied in new, short-lived and durable inputs by means of enhanced quality.

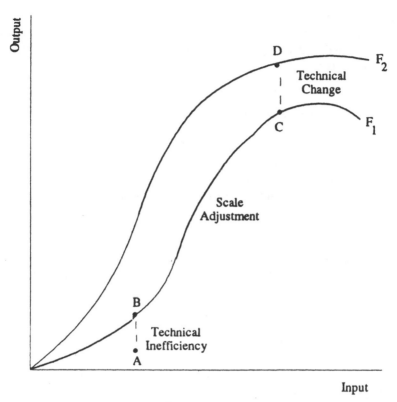

FIGURE 8.1 Decomposition of Productivity Growth into Its Components.

Improved fertilizers, chemicals, seeds and feeds are examples of short-lived (variable) inputs that embody technical change. Similarly, machinery and tractors that perform a greater array of tasks more rapidly, and animal stocks with improved genetic pool are examples of durable (capital) inputs that embody technical change. Common to all these examples is that the innovations are traded in markets and realizing technical change requires investment in the new and improved inputs.

Technical and scale efficiency, on the other hand, result from improvements in the productivity of existing rather than new resources. Improvements in technical efficiency occur when wasteful input use is reduced or when output per unit of input is increased through improved management. Better husbandry, meticulous record keeping, and closer supervision of hired labor and other services are activities that can lead to improvements in technical efficiency in agricultural production. Such activities, however, imply increased managerial effort. Scale efficiencies are realized by re-allocating existing resources to a smaller (greater) number of larger (smaller) production units within a region of increasing (decreasing) returns to scale. Similar to technical efficiencies, scale efficiencies are likely to require additional managerial effort in identifying optimal firm size.

Firm Behavior Under Price Protection

The essence of the foregoing discussion is that agricultural productivity growth is a complex phenomenon resulting from a variety of sources, some of which are generated within the firm while others are generated outside. Productivity growth generated within the firm tends to require greater amounts of managerial effort and better quality management, while productivity growth emanating from innovations traded in the market commands investment. In this section, a behavioral model is presented where the influence of protectionism on the allocation of managerial effort and investment are analyzed for a competitive firm. Initially, the firm is assumed to employ only variable inputs in its production. Subsequently, capital inputs are incorporated in the analysis.

Technical Change Through Variable Inputs

Let the competitive firm be faced in period t with a production technology represented by:

$$Q_t = G(\alpha_t X_t) \, F(M_t, D) \tag{1}$$

where $F(\bullet)$ is assumed to be increasing in its arguments and $G(\bullet)$ satisfies the usual properties of a well behaved production function. X_t is a vector of hired labor and material inputs, and α_t is a vector of efficiency coefficients in period t. This specification implies that technological change is of the embodied variety and X_t are vintage inputs. D represents managerial ability and M_t denotes managerial effort in period t. Managerial ability D is a stock variable, and is determined by an individual's personal traits and talents. M_t is the managerial effort (and time) allocated by the owner-manager among several productive activities. For a given set of inputs X_t, specification (1) implies that managerial ability and effort augment production without affecting the substitution possibilities of inputs.

Within the outlined framework, the effort of the owner-manager is assumed to entail accumulation of information regarding improved inputs, search for means of reducing wasteful use of inputs, improvement in input productivity through better management techniques, and exploration of the possibilities for firm growth to sizes where unit costs of production are minimum. Managerial ability D and the amount of effort M_t allocated by the owner-manager determine the firm's technical and scale efficiency. On the other hand, the level of technical change is determined by the investment in inputs X. As the firm invests in newer and more productive input vintages, technical change follows.

The owner-manager is assumed to maximize utility as a function of income π and leisure L. Income and leisure preferences of the owner-manager are represented by a well behaved utility function U. Under these conditions the level of utility of the owner-manager of the firm in period t is given by:

$$U_t = U(\pi_t, L_t) \tag{2}$$

Leisure is the residual of the manager's time after managerial effort M has been accounted for. Hence, leisure is equal to L=T-M, where T represents the maximum time that the manager could work. Income π equals the rents to management and ownership and hence, the owner-manager's budget constraint in period t is defined by:

$$\pi_t = p_t \, G((\alpha_t X_t) F(M_t, D) - w_t \cdot X_t \tag{3}$$

where w_t is a vector of input prices and p_t denotes output price in period t. To simplify the notation, vintage inputs X_t can be measured in efficiency units and hence re-defined as:

$$V_t = \alpha_t X_t \tag{4}$$

Maximization of the utility function U in (2), subject to the budget constraint (3) and the technological constraint (1), yields the optimum levels of leisure L^*, managerial effort M^*, and investment in variable inputs V^*. In the absence of costs of adjustment or intertemporal changes in preferences, the owner-manager maximizes utility over a single period. After suppressing the time argument for simplicity in notation, the first order conditions of the utility maximization problem are:

$$U_\pi = -\lambda \tag{5a}$$

$$U_M = \lambda p G(V) F_M(M, D) \tag{5b}$$

$$p G_V(V) F(M, D) = w \tag{5c}$$

$$\pi - p G(V) F(M, D) + w \cdot V = 0 \tag{5d}$$

where $U_\pi = \partial U / \partial \pi$, $F_M = \partial F / \partial M$, etc. Conditions (5a) and (5b) taken together imply that, at the optimum, the marginal rate of substitution between managerial effort and income is equal to the marginal value product of managerial effort. Similarly, condition (5c) is the standard input marginality condition where the marginal value products of inputs V are equal to their marginal costs.

Conditions (5a) through (5d) can be used to solve for the optimal levels of managerial effort M^* and investment V^* which individually determine the level of productive efficiency and technical change for the competitive firm. The question is how managerial effort M^* and investment V^*, and hence the firm's productivity levels, change in response to a change in price p due to changes in price protection. The effect of a change in output price p on the effort M^* allocated by the owner-manager and investment V^* can be estimated through comparative statics obtained by totally differentiating (5a) through (5d). The pertinent results are:

$$\partial M / \partial p = \left\{ -U_\pi GF_M \frac{D_{22}}{|D|} \right\} + \left\{ -U_\pi G_V F \frac{D_{32}}{|D|} + FG \frac{D_{42}}{|D|} \right\} \qquad (6)$$

and,

$$\partial V / \partial p = \left\{ -U_\pi GF_M \frac{D_{23}}{|D|} \right\} + \left\{ -U_\pi G_V F \frac{D_{33}}{|D|} + FG \frac{D_{43}}{|D|} \right\} \qquad (7)$$

where | D | is the determinant of the matrix:

$$
\begin{bmatrix}
U_{\pi\pi} & U_{\pi M} & 0 & 1 \\
U_{M\pi} & (U_{MM} + U_\pi p F_{MM} G) & (U_\pi p F_M G_V) & -p F_M G \\
0 & (U_\pi p F_M G_V) & U_\pi p F G_{VV} & 0 \\
1 & -p F_M G & 0 & 0
\end{bmatrix}
$$

and D_{ij} is the cofactor of row i and column j. If the utility function U, the production function $Q = F(\bullet)G(\bullet)$, and the set of input and output prices were known, the magnitude and direction of the effect of protectionism on managerial effort and investment could be directly evaluated through equations (6) and (7). On the basis of general assumptions pertaining to these functions, however, only the general direction of these effects can be examined.

Equation (6) indicates that the change in managerial effort M under a price change is ambiguous. For an increase in output price, the opportunity cost of leisure increases and the firm tends to substitute income for leisure by increasing its supply of managerial effort M. This "substitution" effect is captured by the first bracketed term in (6) which is positive. On the other hand, when output price p increases, the firm can decrease its managerial effort and increase its leisure without sacrificing any of its income π. Such an "income" effect induces the firm to reduce its supply of managerial effort M. This effect is represented by the second bracketed term in (6) and is negative. Hence, when output price increases due to an increase in protection, managerial effort, and hence technical and scale efficiency, increases or decreases depending on the relative magnitude of the income and substitution effects.

The effect of a price change on investment V is more straight forward. As equation (7) indicates $\partial V / \partial p$ is positive, suggesting that an increase (decrease) in output price p induces an expansion (contraction) in investment and hence in technical change. Therefore, the positive effect on technical change generated outside the firm from higher prices due to protectionism that had been hypothesized in some previous studies is verified in this study. However, while the effect of protection on exogenous technical change is clear, its influence on technical and scale efficiency is less transparent from theoretical considerations alone. As such, the overall effect of protectionism on productivity growth is, *a priori*, ambiguous.

Despite the ambiguous influence of protectionism on productivity growth, some general observations on the theoretical relationships derived above could provide useful insights. In particular, it should be noted that the income effect in equation (6) becomes larger as income and output price increase. Hence, for low price and income levels the substitution effect in (6) will tend to dominate the income effect, while for high income and prices the income effect could become dominant. It may be concluded then, that, for low price and income levels, $\partial M / \partial p$ and $\partial V / \partial p$ are both positive and hence a marginal increase in price protection could generate a positive effect on productivity growth through an increase in both technological change and the efficiency of the firm. At high price and income levels, an increase in protectionism could encourage technical and/or scale inefficiencies which may be large enough to overwhelm technical change and result in lower levels of productivity growth.

Technical Change Through Capital Inputs

The results derived above can be generalized further to include capital inputs whose investment flows tend to be lumpy. The argument advanced in this section is that when investment in capital inputs is lumpy, it is possible that increases in output price may generate no new investment by the firm. In fact, depending on the relative size of the pertinent adjustment costs, the vintage and volume of the capital in place, and the level of relative prices, large price changes may be necessary before any investment or disinvestment takes place. This, in effect, implies that price increases due to price protection may be inadequate to generate technical change through investment in new capital inputs. Instead, income effects as described above could induce inefficiencies with a resulting decrease in productivity growth.

The above argument is demonstrated by modifying appropriately the model from the previous section. In particular, the production technology (1) of the firm is now modified to be:

$$Q_t = G(K_t, V_t)\ F(M_t, D) \tag{8}$$

where K_t represents capital inputs expressed in efficiency-equivalent units. Any change in the stock of these capital inputs is assumed to involve costs of adjustments $C(I)$, for which the following conditions are assumed to hold:

$$C(0) = 0 \tag{9a}$$

$$C(I_t) > 0 \text{ for } I_t \neq 0 \tag{9b}$$

$$C'(I_t) \text{ exists for all } I_t \text{ other than } I_t = 0 \tag{9c}$$

$$\lim_{I \to 0^-} C'(I_t) \neq \lim_{I \to 0^+} C'(I_t) \text{ but both exist} \tag{9d}$$

where I_t denotes gross investment in period t. Assumptions (9a) and (9b) imply that $C(I)$ has a minimum at point $I = 0$ while assumptions (9c) and (9d) suggest that $C'(I)$ is discontinuous at point $I = 0$. The discontinuity of $C'(I)$ implies that at the neighborhood of $I = 0$, adjustment costs associated with investment, such as installation and delivery costs, cannot be recovered on sale. Similarly, adjustment costs of disinvestment, such as costs of detaching and moving capital, cannot be recovered on purchase. Such costs of adjustment have been used to explain lumpy investment behavior (Rothchild 1971; Hsu and Chang 1990).

The owner-manager is, as before, assumed to maximize utility as a function of income and leisure. However, due to the presence of adjustment costs in capital investment, the maximization is now a multi-period rather than a single period problem. The intertemporal utility maximization problem may be expressed as:

$$\max \sum_{t=1}^{\infty} [U(\pi_t, M_t) - D(I_t)] (1+r)^{-t}$$

s.t. $$\pi_t = p_t G((K_t, V_t) F(M_t, D) - u_t \cdot K_t - w_t \cdot V_t \tag{10}$$

$$K_t = K_{t-1} + I_t$$

where r is the discount rate and u_t is the price vector of capital inputs at period t. D(I) is defined as the disutility that results from the loss of profits due to the costs of adjustment C(I) associated with investment I. In addition, D(I) is assumed to capture the psychological/emotional costs that may result from adjusting the capital stock "too fast", either upwards or downwards, in the face of uncertainty. Risk averse entrepreneurs may adopt a "wait and see" approach in capital investment/divestment due to the fear of over- or under-capitalizing. Hence, the anxiety of uncertainty in investment decisions is captured by D(I). Additional psychological costs may also be present when downsizing due to a sense of loss in net worth or wealth. Given that D(I) is proportional to C(I), conditions parallel to (9) are assumed to hold for D(I),

$D(0)=0$

$D(I_t)>0$ for $I_t \neq 0$

$D'(I_t)$ exists for all I_t other than $I_t=0$ (11)

$$\lim_{I \to 0^-} D'(I_t) \neq \lim_{I \to 0^+} D'(I_t) \text{ but both exist.}$$

Hence, D(I) is assumed to have a minimum at I=0 and D'(I) to be discontinuous at the neighborhood of I=0. The reasoning is straightforward. If the costs of adjustment associated with investment (disinvestment) C(I) are discontinuous at the neighborhood of I=0, the disutility function D(I) which is proportional to C(I) will also be subject to similar discontinuities. Now let $U^*(K,p,w,u)$ be the outcome of the optimization:

$$U^*(K,p,w,u) = \max U(\pi,M)$$

s.t. $\pi = p\, G((K,V)F(M,D) - u \cdot K - w \cdot V.$ (12)

Hence, $U^*(K,p,w,u)$ is the maximum attainable utility from having capital stock K available in period t.   Under this condition, the utility maximization problem of the firm in (10) can be rewritten as an infinite horizon problem:

$$\max \Re(K_0,I,p,w) = \sum_{t=1}^{\infty} [U^*(K_t,p,w) - D(I_t)](1+r)^{-t}$$

s.t. $K_t = K_{t-1} + I_t$
$$K(0) = K_0$$ (13)

In essence, the above optimization procedure implies a two step decision rule. In the first step, the entrepreneur decides the level of capital investment while the second step determines the optimal levels of variable inputs and managerial effort to be devoted to the capital stock at hand.

Because of the discontinuity of the D(I) function at point I=0, the optimization problem (13) may be mathematically restated, after suppressing p and w for simplicity, as:

$$\begin{cases} \max\Re(K_0,I) \text{ s.t. } I \geq 0 \\ \max\Re(K_0,I) \text{ s.t. } I \leq 0 \end{cases}$$ (14)

Due to the presence of the inequalities in (14), the Kuhn-Tucker conditions are used to guide the derivation of the optimal investment program. If I^* denotes the optimal investment sequence over the planning horizon, the optimality conditions suggest that:

$$\begin{cases} \max\Re'(K_0,I^*) \text{ s.t. } I \geq 0 \text{ and } I^* \cdot \Re'(K_0,I^*) = 0 \text{ for } I^* \geq 0 \\ \max\Re'(K_0,I^*) \text{ s.t. } I \leq 0 \text{ and } I^* \cdot \Re'(K_0,I^*) = 0 \text{ for } I^* \leq 0 \end{cases}$$
                                                                            (15)

where $\Re'(K_0,I^*)$ is the derivative of $\Re$ with respect to investment I evaluated at I^*. Condition (15) suggests that the entrepreneur is faced with the dual decision of whether or not to invest and at what level. Combining (13) and (15), the decision rule for optimal investment at period 1 is further clarified. In the case where:

$$\lim_{I_1 \to 0^-} D'(I) < \sum_t [U^{*'}/(K_0, I_1^*, I_2^*, ...)](1r)^{-t} < \lim_{I_1 \to 0^+} D' v(I) \tag{16}$$

holds, no adjustment in the capital stock of the firm takes place. If, on the other hand, the present value of the marginal utility from an additional unit of capital lies outside the upper/lower limits of D'(I), then investment/disinvestment takes place in period 1 at a level that satisfies the marginal rule:

$$\sum_{t=1}^{\infty} [U^{*'}(K_0, I_1^*, I_2^*, ...)](1+r)^{-t} = D'(I_1^*). \tag{17}$$

Both such decisions are illustrated in Figure 8.2. Point A represents the lim D'(I) for I→0$^+$ while B represents the lim D'(I) for I→0$^-$ and $\Im$ is the present value of the marginal utility from an additional unit of investment in period 1:

$$\Im = \sum_t {}^{\infty} [U^{*'}(K_0, I_1, I_2, ...)] (1+r)^{-t}.$$

As depicted in Figure 8.2, $\Im$ could fall between points A and B, in which case no adjustment takes place as the existing capital stock is in equilibrium.

It is of interest then to investigate the influence of an output price increase due to government protectionist actions on capital investment and its embodied technical change. As price p increases, *ceteris paribus*, the marginal value product of capital increases and so does the marginal utility from an additional unit of capital. Hence, $\Im$ tends to shift upwards with p as illustrated in Figure 8.2. However, this increase in output price may or may not be sufficient to generate additional investment depending on whether $\Im$ is shifted beyond point A. Thus, depending on the initial position of $\Im$ as well as the level of increase in p, protectionistic actions that raise output prices may or may not generate additional capital investment and hence technical change in agricultural production.

The role of the initial stock K_0 is also illustrated in Figure 8.2. Larger initial stocks K_0 tend to decrease the marginal productivity of additional units of capital and hence the marginal utility resulting from an

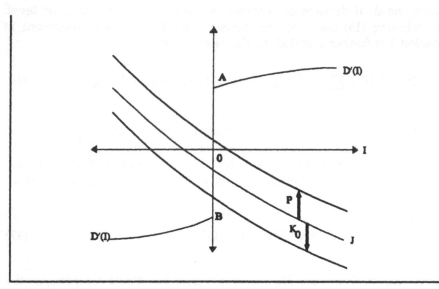

FIGURE 8.2 Investment Behavior with Discontinuous Adjustment Costs

additional unit of investment. Thus, as K_0 increases, $\mathfrak{I}$ shifts downwards. As a result, large price increases may be necessary to generate additional investment and technical change for agricultural industries with large and relatively updated capital stocks.

As soon as the decision on capital investment I is made, the entrepreneur can choose the optimal levels of variable inputs V_t and managerial effort M_t conditional on the capital stock available to the firm. For these decisions the results derived in equations (5) through (7) still hold. Hence, the results of the two models can now be put together. In cases where technical change in agricultural production is primarily embodied in variable inputs, protectionist actions will tend to positively influence the investment and hence technical change in production. In cases where technical innovations are embodied in capital inputs, price protectionism, *a priori*, has an ambiguous effect on investment and technical change. If the initial capital stock of the protected industry is low in volume and/or obsolete, a large enough increase in output price from protectionistic actions could generate new investment and technical change. If on the other hand the initial capital stock of the protected industry is large in volume and relatively updated, then even large increases in prices may not generate additional investment and technical change. The effect of price protection on managerial effort, and hence productivity growth that results from increases in technical and scale efficiencies as well as technical change generated within the agricultural firm is also, *a priori*, ambiguous. For low income industries the effect of price protection on managerial effort could be positive while for high income industries income effects are likely to generate technical and scale inefficiencies.

From the behavioral models presented above, the influence of government price-protectionism on productivity growth is found to be, *a priori*, ambiguous. A positive influence of protectionism on productivity growth, as postulated by previous studies, need not hold. Price and income levels, existing capital stock, and the relative size of the protectionist price-hike all tend to influence the impact that government protectionistic actions have on agricultural productivity growth. Ultimately, whether protectionism has a positive or a negative effect on agricultural productivity growth is an empirical issue.

A Relevant Empirical Model

In the previous section, technical change as well as scale and technical efficiency were considered *endogenous* in that their levels were decided by the firm at any given point of time. Hence, an empirical model that is consistent with these assumptions, should allow for endogenous technical change and efficiency. Most existing empirical models of productivity growth consider technical change to be exogenous to the firm. Few attempts exist in the literature where agricultural technical change and productivity growth are allowed to be endogenously determined by the firm. A notable exception is the recent work by Mundlak (1988). In particular, Mundlak (1988) considered an aggregate production function with endogenous technical change by specifying optimal output to be:

$$F(x^*,s) = y^*$$
(18)

where y^* is the maximum output in a given production process, x^* is the optimal input levels, and s is a vector of *state variables* which determine the level of technical change. State variables are exogenous to the firm. The entrepreneur is assumed to observe such variables and choose the technology to be used in the production process. Hence, technical change is endogenous to the firm in the sense that it is subject to choice.

Recognizing that the firm may diverge from x^* due to inefficiencies, Mundlak (1988) suggested that $F(x^*,s)$ be approximated to the second order around x^* by a function g(x,s) as:

$$g(x,s) = F(x^*,s) + (x-x^*)' \triangledown F(x^*,s) + \frac{1}{2}(x-x^*)' \triangledown^2 F(x^*,s)(x-x^*)$$

$$= \Gamma(x^*,s) + B(x^*,x,s)$$
(19)

here:

$$\Gamma(x^*,s) = F(x^*,s) - x^{*\prime}\nabla F(x^*,s) + \tfrac{1}{2}x^{*\prime}\nabla^2 F(x^*,s)\, x^*$$

$$B(x^*,x,s) = \nabla F(x^*,s) - \nabla^2 F(x^*,s)x^* + \tfrac{1}{2}\nabla^2 F(x^*,s)x.$$

Estimation of $g(s,x)$ requires estimation of $\Gamma(x^*,s)$ and $B(x^*,x,s)$ which are unknown functions in s and the unobserved x^*. As Mundlak (1988) pointed out, however, it is not possible to determine x^* separately from B and Γ. Instead, a reduced form can be obtained by expanding these functions around s as:

$$\Gamma(x^*,s) \approx \pi_0 + s'\pi_{10} + s'\pi_{20}s$$

$$B(x^*,x,s) \approx \pi_{01} + \pi_{s1}s + \pi_{x1}x \tag{20}$$

where π_0 is a constant, π_{10} and π_{01} are vectors, and $\pi_{20}, \pi_{s1}, \pi_{x1}$ are matrices conformably specified. Combining (19) and (20), the final form of the production function $g(x,s)$ is:

$$g(s,x) \approx \pi_0 + s'\pi_{10} + s'\pi_{20}s + x'\pi_{01} + x'\pi_{s1}s + x'\pi_{x1}x. \tag{21}$$

The above specification of a production function allows for endogenous technical change to be determined by a vector of state variables s. Hence, it is an improvement over traditional production functions with exogenous technical change. However, it is also restrictive in that it does not allow for technical and scale efficiencies to be endogenously determined. For that reason, the production function in (21) is generalized in this study to allow for endogenously determined levels of efficiency. Furthermore, the restrictive assumption of a single period planning horizon imposed by Mundlak (1988) in his analysis is also relaxed.

Assuming that technical change as well as technical and scale efficiency are endogenously determined relative to an appropriate vector of state variables s, maximum output of the ith firm in period t for the technology specified in (1) is given by:

$$y_{it}^* = G(X_{it}^*, s_{it})F(M_{it}^*, s_{it}; D) \tag{22}$$

where X_{it}^* denotes the optimum levels of capital and variable inputs employed by the ith firm and M_{it}^* is the optimal managerial effort. Allowing the levels of capital and variable inputs as well as managerial effort to diverge from their optimal levels, the production function in (22) can be approximated by a function $g(s,X)$ following procedures similar to those in the derivation of (21). Specifically, approximating (22) to the second order around X^* yields:

$$g(x,s) = \Delta(X^*,M^*,s) + H(M^*,X^*,X,s)$$

where $\Delta(X^*,M^*,s) = \Gamma(X^*,s)F(M^*,s;D)$

and

$$H(M^*,X^*,X,s) = B(X^*,X,s)F(M^*,s;D) \tag{23}$$

Expanding $\Delta(X^*,M^*,s)$ and $H(M^*,X^*,X,s)$ around s in a fashion similar to that in (20), and combining with (23) yields the final form of the production function:

$$g(s,X) = \delta_0 + s'\delta_{10} + s'\delta_{20}s + X'\delta_{01} + X'\delta_{s1}s + X'\delta_{x1}X. \tag{24}$$

The primary difference of the production function in (24) from that in (21) is that the components δ_0, $s'\delta_{10}$, $s'\delta_{20}s$, vary both with time and across firms as the managerial effort and abilities vary from one firm to another, and, over time. Hence, the sum $(\delta_0 + s'\delta_{10} + s'\delta_{20}s)$ can be interpreted as time-varying technical and scale efficiency effect which is endogenously determined by an appropriate set of state variables.

An Empirical Test: The Case of
New Zealand Beef/Sheep Industry

The insights gained through the theoretical developments are tested empirically in this section within the framework of the New Zealand beef/sheep industry which enjoyed substantial government protection during the 1960s, 1970s, and early 1980s. In 1985, the protectionist policies were removed. Since then the industry has been undergoing substantial adjustment. However, the number of firms in the industry has remained stable since 1985. Market liberalization did not cause substantial exodus of small or inefficient farms. Hence, the reaction to the liberalization as manifested in secondary data is likely to be reflective of all the farms in the industry rather than just of the most efficient ones. Thus, due to this experience, the New Zealand beef/sheep industry provides a unique economic experiment whereby the influence of protectionism on productivity growth can be investigated.

The influence of protectionism on productivity growth of the New Zealand beef/sheep industry is tested empirically within the framework of a production function with endogenous technical change and efficiency. In particular, the following restricted form of the production function in (23) is specified and estimated:

$$y_{it} = g(s_1, s_2, x) = \alpha_{i0} + s_{1it}'\alpha_{10} + s_{1it}'\alpha_{20}s_{1it} + x_{it}'\beta_1 + x_{it}'\beta_2 s_{2it} \qquad (25)$$

where $x = (LAB, ND, MT, BK, MC)$

$s_1 = (INC, W, PROT)$

$s_2 = (K_{-1}, T, PROT)$

Input vector x includes labor (LAB), land and improvements (ND), material inputs (MT), breeding stock (BK), and machinery (MC). LAB is the aggregate of hired permanent and seasonal labor hours as well as contract labor hours. ND is measured in constant (1976) value. Although acreage is also available, substantial differences in the type of terrain among different regions exist. Such differences reflect both the productive capacity and its value. To account for land heterogeneity, the value of land and improvements in constant NZ dollars rather than total acreage was used. MT is the aggregate of chemical inputs, fertilizer, as well as fuel and lubricants. BK is measured by the number of breeding animals at the beginning of the year. Finally, MC is measured by the real market value of machinery in 1976 dollars. Output and inputs are specified in logarithmic terms. Input-output information from eight production regions over the period 1975-76 to 1990-91 is used for the estimation of (25). Thus, i indexes regions of production (i=1,..,8) and t indexes time (t=1,..,16) in (25).

Vector s_1 includes the state variables that are assumed to influence technical and scale efficiency, namely, the level of income (INC) and the level of protection (PROT). INC is measured as net real income from farming in period t-1 while PROT is measured by producer subsidy equivalents (PSEs) for the New Zealand pastoral agriculture. The total "efficiency" effect for each region i is captured by the sum $(\alpha_{i0} + s_{1it}'\alpha_{10} + s_{1it}'\alpha_{20}s_{1it})$. A weather variable (W) is also included in the state vector s_1 to capture possible weather effects on output. W is specified as the total number of days of soil moisture deficit in a year for each producing region. This index measures the number of days when there is insufficient moisture in the soil to allow pasture growth.

Vector s_2 consists of state variables that are assumed to affect the investment of variable and capital inputs, namely, the level of protection (PROT), and the capital stock already in place K_{-1}. In addition, T has been included to capture the availability of improved technology over time. PROT is, as previously, measured by PSEs for the New Zealand pastoral agriculture. K_{-1} is the stock of all capital inputs in period t-1. That is, K_{-1}

is an aggregate of land and improvements, machinery, and breeding stock in period t-1. Finally, T is a trend variable.

The production function specified in (25), along with the value share equations of labor, materials, machinery, and breeding stock were estimated through iterative three stages least squares procedures and the parameter estimates are given in Table 8.1. Overall, the statistical fit of the estimated model is deemed adequate as twenty out of a total of thirty one estimated parameters were statistically significant at the 5 percent level.

The results indicate that statistically significant differences in efficiency among regions exist as captured by the separate intercepts R_i (i=2,..8). Out of the eight production regions, high country farms in South Island (region 1) were found to be the least efficient, while intensive finishing farms in South Island (region 8) were the most efficient in beef/sheep production. Variations in weather, as captured by W, were found to have statistically insignificant effects on beef/sheep production.

The effects of real farm income INC and INC^2 on the "efficiency" of all regions were both found to be negative but statistically insignificant. Thus, existing differences in real income among the eight producing regions are not sufficient to generate any differences in "efficiency." The coefficients associated with the level and rate of protection, however, were both statistically significant and negative. Thus, over the period of analysis, "inefficiency" in New Zealand beef/sheep production increased at an increasing rate with protection. This result is in agreement with the conclusions obtained from the theoretical model. For a high income industry, such as the New Zealand beef/sheep industry, protection translates into an income effect and tends to reduce the efficiency level of production.

The effects of protectionism on input investment are also in agreement with the theoretical results obtained earlier. In particular, the level of investment in land and improvements, intermediate inputs, and machinery in beef/sheep production increased with the level of protection over the period of interest. Investment in labor and breeding stock were unaffected by the level of protection. Given that the New Zealand government focused its input subsidization in land development programs, subsidies to fertilizers, and tax write-offs for capital investment, the results here indicate that such measures were successful in generating positive investment response. However, the levels of growth in the productivity of each of these inputs from increased investment due to protectionism were modest: 0.6 percent per year increase in land productivity, 0.2 percent increase in the productivity of intermediate inputs, and 0.06 percent increase in the productivity of machinery for each additional unit of protection.

TABLE 8.1 Production Function Parameter Estimates for New Zealand Beef/Sheep Industry

Parameter	Estimated Value	t-Statistic
a_0	0.2229	0.3804
R_2	0.1588	5.1914*
R_3	0.3038	4.6388*
R_4	0.4342	6.4253*
R_5	0.5040	10.182*
R_6	0.3883	7.1344*
R_7	0.5416	6.4077*
R_8	0.5572	27.843*
a_W	-0.0001	-0.5856
a_{Inc}	-0.0015	-0.4404
a_{Inc2}	-0.0002	-0.6770
a_{Prot}	-0.0357	-3.8068*
a_{Prot2}	-0.0004	-7.0011*
b_{Lab}	0.1494	9.5948*
b_{Nd}	0.2775	2.5704*
b_{Mat}	0.2427	14.887*
b_{Bk}	0.1409	8.2258*
b_{Mc}	0.0314	4.0841*
$b_{Lab,Pr}$	0.0007	1.3974
$b_{Lab,T}$	0.0003	0.3053
$b_{Nd,Pr}$	0.0066	3.6903*
$b_{Nd,T}$	-0.0033	-1.7837
$b_{Nd,K-1}$	-0.0015	-0.2905
$b_{Mat,Pr}$	0.0024	4.6698*
$b_{Mat,T}$	0.0035	3.4543*
$b_{Bk,Pr}$	-0.0010	-1.5007
$b_{Bk,T}$	0.0015	0.8480
$b_{Bk,K-1}$	0.0077	1.1211
$b_{Mc,Pr}$	0.0006	1.9806*
$b_{Mc,T}$	0.0039	5.3975*
$b_{Mc,K-1}$	-0.0081	-3.1689*

Note: Adj R-squared = 0.67.
             DW statistic = 1.59.
* Indicates statistical significance at the 5 percent level.

The total effect of protectionism on productivity growth in New Zealand beef/sheep production can be evaluated by differentiating y_{it} with respect to PROT. This effect is the sum of the "technical change" and the "efficiency" effects. $\partial y_{it}/\partial PROT_t$ was evaluated at each sample point. For the period 1974-75 to 1984-85, the overall effect of protection averaged over all regions of production was slightly negative. While an average annual increase in output of almost 5 percent was generated from technical change, the average effect from inefficiencies induced by the protectionistic environment was -5.5 percent. As a result, productivity growth decreased by 0.5 percent per year due to protection over the aforementioned period. For the period 1985-86 to 1990-91 where protectionism was substantially reduced, improvements in efficiency were realized and as a result the overall effect of protectionism on productivity growth was positive and equal to 1.5 percent per year. Hence, the experiences of the New Zealand beef/sheep industry corroborate the hypothesized effects of protectionism on productivity growth.

The amount of existing capital stock was found to have a statistically significant negative effect only on machinery investment. Investment in land and improvements and breeding stock appear to have been unaffected by the amount of existing capital stock. Finally, availability of technology, as captured by the trend variable T, was found to have statistically significant and positive effects only on material and machinery inputs. In particular, the assumed availability of technology contributed an annual increase of 0.3 percent in the productivity of material inputs and almost 0.4 percent in the productivity of machinery.

Summary and Conclusions

The relationship between protection and productivity growth has been investigated in this study. This relationship was first analyzed within a theoretical model for a competitive firm. It was concluded that the effect of protectionism on productivity growth is *a priori* ambiguous. This result should be contrasted with the positive effect that protectionism was assumed to have on productivity growth in previous studies. Based on reasonable assumptions, however, it was concluded that for firms with low income and low volume of capital stock a marginal increase in protectionism may have a positive effect on productivity growth by encouraging investment and technical change. For firms with high incomes and large capital stock already in place, protectionism tends to generate technical and scale inefficiencies which could substantially reduce productivity growth.

Empirical evidence from the New Zealand beef/sheep industry supported the hypothesis that high levels of protectionism for high income industries is likely to be negatively related with efficiency and productivity growth. Using a production function with endogenous technical change and efficiency levels, it was concluded that high levels of protection during the 1975-1985 period produced negative productivity growth as technical change encouraged by price-protection was overwhelmed by inefficiencies related to the protectionistic environment. Following the liberalization of the market in 1985, such inefficiencies decreased substantially and as a result productivity growth was encouraged. The results of this study corroborate similar results in Kalaitzandonakes and Taylor (1990), and Kalaitzandonakes, Gerhke, and Bredahl (1994), where intensified competition was found to have sustainable positive effects on productive growth in high income industries.

The results of this study call into question analyses of market and trade liberalization that do not explicitly consider adjustments due to increases in efficiency and productivity growth. In the case where the results presented here based on an explicit firm-level model and commodity-specific analysis are more generally realized, the cost of protectionism and, correspondingly, the benefits of liberalization have been significantly underestimated.

Notes

1. The homothetically separable form of the production function in (1) follows from the basic premise that managerial effort and abilities act to improve the overall efficiency of production without influencing the substitution possibilities of inputs. This is in agreement with the usual Farrell-type measures of technical efficiency that exist in the literature.

2. This outcome is in agreement with the empirical results of Fulginiti and Perrin (1994) who found that agricultural sectors in developing countries, usually characterized by low prices and income, benefited from protectionism.

3. Throughout the theoretical developments in this study, the process of scientific discovery has been largely ignored. Implicitly, the existence of a continuum of new and improved technologies has been assumed. Including T in the state vector s_2 is consistent with this assumption. However, appropriate indexes of scientific discovery that capture shifts or discrete jumps in availability of new technology may be more appropriate, when available.

4. Over the period of analysis, the elasticities of scale for all eight regions of production were found to be close to 1. Approximately constant returns to scale suggest that the likelihood of scale inefficiencies are small. Hence, most of the "inefficiency" effect in beef/sheep production is likely to represent technical inefficiency.

References

Antle, J.M. 1988. "Dynamics, Causality, and Agricultural Productivity," in S.M. Capalbo and J.M. Antle, eds., *Agricultural Productivity Measurement and Explanation*. Washington D.C.: Resources for the Future.

Binswanger, H.P. 1974. "The Measurement of Technical Change Biases with Many Factors of Production." *American Economic Review* 64:964-76.

Farrel, K., and T. Runge. 1983. "Institutional Innovation and Technical Change in American Agriculture: The New Deal." *American Journal of Agricultural Economics* 65:1168-73.

Fulginiti, L.E. and R.K. Perrin. 1993. "Prices and Productivity in Agriculture." *Review of Economics and Statistics* 75:471-482.

Hayami, Y. and V.W. Ruttan. 1985. *Agricultural Development*. Baltimore, MD: John Hopkins University Press.

Hicks, J.R. 1963. *The Theory of Wages*, 2nd Edition. London: MacMillan.

Hsu, S., and C. Chang. 1990. "An Adjustment-Cost Rationalization of Asset Fixity." *American Journal of Agricultural Economics* 72:298-308.

Huffman, W. and R. Evenson. 1992. *Science for Agriculture*. Ames, IA: Iowa State University Press.

Kalaitzandonakes, N.G., and T.G. Taylor. 1990. "Competitive Pressure and Productivity Growth: The Case of the Florida Vegetable Industry." *Southern Journal of Agricultural Economics* 22:13-21.

Kalaitzandonakes, N.G., B. Gerhke, and M.E. Bredahl. 1994. "Competitive Pressure, Productivity Growth and Competitiveness," in M. Bredahl, P. Abbott, and M. Reed, eds., *Competitiveness in International Food Markets*. Boulder, CO: Westiview Press.

Koester, U. 1991. "The Experience with Liberalization Policies: The Case of the Agricultural Sector." *European Economic Review* 35:562-570.

Leibenstein H. 1966. "Allocative Efficiency Versus X-efficiency." *Amerian Economic Review* 56:392-415.

Mundlak, Y. 1988. "Endogenous Technical Change and the Measurement of Productivity," in S.M. Capalbo and J.M. Antle, eds., *Agricultural Productivity Measurement and Explanation*. Washington, D.C.: Resources for the Future.

New Zealand Meat and Wool Boards' Economic Service. 1975-1991. *The New Zealand Sheep and Beef Survey*. Wellington, various issues.

_____. 1975-1991. *Annual Review of the New Zealand Beef and Sheep Industry*. Wellington, various issues.

Nishimizou, M. and J. Page. 1982. "Total Factor Productivity Growth, Technological Progress, and Technical Efficiency Change: Dimensions of Productivity Change in Yugoslavia, 1965-78." *Economic Journal* 92:920-36.

Reynolds, R. and S. SriRamaratnam. 1990. "How Farmers Responded," in R. Sandrey and R. Reynolds, eds., *Farming Without Subsidies*. New Zealand Ministry of Agriculture and Fisheries.

Robinson, J. 1956. *The Accumulation of Capital*, Homewood, IL: Irwin.

Rothchild, M. 1971. "On the Costs of Adjustment." *Quarterly Journal of Economics* 85:605-22.

Schultz, T.W., ed. 1979. *Distortions of Agricultural Incentives.* Bloomington, IN: Indiana University Press.

Tyler, L. and R. Lattimore. 1990. "Assistance to Agriculture," in R. Sandrey and R. Reynolds, eds., *Farming Without Subsidies.* New Zealand Ministry of Agriculture and Fisheries.

VanderMeer, C.L. and S. Yamada. 1990. *Japanese Agriculture: A Comparative Economic Analysis.* London: Routledge.

9

Detailed Patterns of Intra-industry Trade in Processed Food

Joseph G. Hirschberg and James R. Dayton

Introduction

A significant portion of world trade in processed food is increasingly composed of intra-industry trade (IIT), where IIT is trade between two countries in products that are close substitutes for each other (Tharakan 1985). A widely studied example of IIT is in manufactured goods, such as automobiles. Unlike the neoclassical theory of comparative advantage, where nations with complementary resources trade, Linder (1961) has argued that IIT will be most prevalent between countries with similar economies.

Recently a number of models have been developed to explain IIT based on imperfect competition, where economies of scale and imperfect competition are the most prevalent factors contributing to the development of intra-industry trade in a particular industry. These models have been developed in a series of works by Krugman (1979, 1981), Dixit and Norman (1980), Helpman (1981), and Helpman and Krugman (1985).

Empirical studies of intra-industry have been conducted on a variety of industries. Greenaway and Milner (1986), Balassa and Bauwens (1987), Helpman (1987), and Bergstrand (1990) investigate the determinants of IIT in the aggregate and for industrial sectors. Also, recent studies that investigate IIT in the processed food industry have been conducted by McCorriston and Sheldon (1991), Hart and McDonald (1992), Christodoulou (1992), and Hirschberg, Sheldon, and Dayton (1994) (hereafter HSD).

In this chapter we follow the form of the regression estimated in HSD to model IIT. Instead of aggregating over all processed food industries

(SIC=20) and fitting one set of parameters, this chapter investigates the patterns of intra-industry trade in processed food by identifying patterns in the regression coefficient values of 49 regression models constructed to explain the levels of IIT in each sector.

The analysis in this chapter uses the UN trade flow data for the period 1964 to 1988 and the Penn World Tables as constructed by Summers and Heston (1991). IIT at the sector level is modelled using the characteristics of the trading partners: intercountry comparisons of capital labor ratios, the comparative sizes of the economies, the distance between partners, long-term fluctuations in exchange rates, common borders, and membership in customs union/free trade zones. This will result in the estimation of 49 separate models based on the modified SIC definitions.[1]

The interpretation of the present results is complicated by the number of parameters estimated (92 per sector). In order to facilitate this analysis a specialized form of cluster analysis is used. This technique clusters the estimated parameters from the IIT models fitted to each 4-digit SIC by incorporating their estimated covariances in the cluster definitions and distances.

By classifying the particular SIC=20 industries by the characteristics of their IIT we can assess the degree to which this trade is a function of the inherent technology and market structure of the sector. This study identifies those industries where IIT is most greatly influenced by geography, trade conventions and national technological characteristics while accounting for country and time-specific factors. These results can be used to determine which industries may benefit the most from policies designed to encourage trade.

Model

Following Helpman and Krugman (1985, pp.169-178) three testable hypotheses can be stated concerning the level of IIT between two countries (see also Helpman 1987 and Wickham and Thompson 1989). First, the level of IIT will be higher (lower) the greater the equality (inequality) of relative factor endowments between countries. As a country's income is a function of its capital-labor ratio in this model, the hypothesis can be re-stated as: the level of intra-industry trade will be higher (lower) the greater the equality (inequality) of countries' GDP per capita. Second, the degree of IIT will be higher (lower), the smaller (greater) the relative size of the capital-rich country, size being measured by GDP. And third, Helpman and Krugman (1985) argue that relatively more differentiated goods will be produced in more capital intensive industries. Thus, the IIT for a specific country will be positively associated with endowments of capital per worker, as measured by a country's per capita income (a result already confirmed by Havrylyshyn and Civan 1984).

The model used in this analysis follows the relationship proposed in HSD and is similar to Noland's (1988) model. The Grubel and Lloyd (1975) measure of intra-industry trade between two countries (IIT) is defined as:

$$IIT_{jki} = 1 - \frac{|X_{jki} - M_{jki}|}{X_{jki} + M_{jki}} \tag{1}$$

where IIT_{jki} indicates intra-industry trade between country j and country k for industry i. X_{jki} is the value of exports for industry i reported by country j as coming from country k, and M_{jki} is the value of imports for industry i reported by country k as coming from country j. The index tends towards one in the case of intra-industry trade and zero for inter-industry trade. In this analysis we assume that IIT is a doubly truncated normally distributed random variable (a Tobit model as defined by Tobin 1958).

The model used in the analysis is defined as:

$$IIT_{jkt} = \alpha_0 + \beta_1 INEQGDC_{jkt} + \beta_2 GDPSIZE_{jkt} + \beta_3 GDC_{kt} + \beta_4 DEX_{jkt}$$

$$+ \beta_5 DIST_{jkt} + \beta_6 DIST2_{jkt} + \beta_7 DIST3_{jkt} + \beta_8 BORDER_{jkt} + \beta_9 EC_{jkt} \tag{2}$$

$$+ \beta_{10} EFTA_{jkt} + \sum_{j=1}^{29} \lambda_j DRC_j + \sum_{k=1}^{29} \pi_k DPC_k + \sum_{t=65}^{88} \theta_t DYR_t + \varepsilon_{jkt}$$

where IIT_{jkt} is the Grubel and Lloyd measure of intra-industry trade between importing country j and exporting country k in year t, the β_i are the estimated coefficients for the variables defined above, DRC_j and λ_j are the dummy variable for reporting country j and the associated coefficient, DPC_k and π_k are the dummy variable for partner country k and the associated coefficient, DYR_t and θ_t are the dummy and associated coefficient for year t, and ε_{jkt} is a normally distributed random variable with a zero mean and a constant variance σ_ε^2. Detailed definitions of the independent variables are given in the remainder of this section.

$INEQGDC_{jk}$ is an indicator of inequality between the importing country's GDP per capita (denoted as GDC_j) and the exporting country's (GDC_k) and is employed here as a proxy for inequality in the capital to labor ratio between both countries and is expected to have a negative effect on IIT. Following Balassa and Bauwens (1987), this is defined as

$$\text{INEQGDC}_{jk} = 1 + \frac{[w_{jk}\ln(w_{jk}) + (1-w_{jk})\ln(1-w_{jk})]}{\ln(2)} \tag{3}$$

where $w_{jk} = \text{GDC}_j / (\text{GDC}_k + \text{GDC}_j)$, and INEQGDC_{jk} varies over the range 0 to 1.

GDPSIZE_{jk} is defined as an index of GDP of the importing country relative to the exporting country. This variable indicates the relative size of the trading partners, given differences in their capital-labor ratios. GDPSIZE_{jk} has non-zero values in cases where the per capita GDP's vary:

$$\text{GDPSIZE}_{jk} = \left(\frac{\text{GDC}_k - \text{GDC}_j}{\text{GDC}_k + \text{GDC}_j}\right)\left(\frac{\text{GDP}_j}{\text{GDP}_k}\right) \tag{4}$$

When per capita incomes are different this implies that capital endowments differ. Thus, the ratio of GDPs should be important in determining the level of IIT_{jk}. Otherwise it should not have an influence. The sign on the parameter for this variable should be positive under the Helpman-Krugman (1985) hypothesis.

The value of GDP per capita for the reporting country (GDC_j) is used to account for the influence of the implied endowment of capital per worker. It is expected that the larger the capital-labor ratio the greater the IIT_{jk}.

To allow for the possible impact of long-run exchange rate volatility, a variable is included that measures the absolute value of the one year proportional change in the exchange rate between the reporting and partner country, (DEX_{jkt}):

$$\text{DEX}_{jkt} = \left(\frac{|\ ex_{jt} - ex_{jt-1}\ |}{|\ ex_{kt} - ex_{kt-1}\ |}\right)\left(\frac{ex_{kt}}{ex_{jt}}\right) \tag{5}$$

where ex_{jt} and ex_{kt} are the dollar exchange rates for country j and k at time t. De Grauwe and de Bellefroid (1987) argue that the appropriate time-frame for studying the effects of exchange rate fluctuation on the volume of trade is at least a year, if not longer. Here we assume, *a priori*, that the implications for IIT_{jk} are similar to their findings for aggregate trade. A factor that negatively influences all bilateral trade should have an equivalent impact on intra-industry trade in processed food.

The effect of the trading partners' relative location is modelled using an estimated cubic function in the distance between geographic centers of each country. These distances were computed using spherical geometry and the latitude and longitude of the geographic center of each country as given in the SAS data set WORLDMAP (SAS Institute Inc. 1989). Dummy variables are included for trade between countries that are either both in the European Community (EC_{jk}), or both in the European Free Trade Association ($EFTA_{jk}$), and to account for trading countries sharing a common land border ($BORDER_{jk}$).

The specification also includes dichotomous variables for each country as an exporter and as an importer to account for country-specific unobserved factors such as tariff and non-tariff barriers which are not accounted for by the variables defined above. Additional dummy variables are also included for each year in order to account for international events such as the energy price shocks and other year specific events.

Data

The trade data used in this study are the reported value of imports in U.S. dollars from the D-series trade data complied by the United Nations. Trade data for the 30 nations which engage in the highest level of trade in the SIC=20 industries for the period from 1964-1988 were combined with the purchasing power parity GDP values from the Penn Mark 5 data base as described in Summers and Heston (1991).

The data from the UN is reported by SITC, which do not correspond to SICs directly. The aggregation of SITCs into 4-digit SICs was based in large part on the aggregation used by Dayton and Henderson (1992), with a number of modifications. In a number of cases, the 4-digit SIC codes were subdivided when it was apparent that the demand for the good may be quite different. For example if the definition of SIC=2011 is used to define trade, trade in both meat packing products and undressed hides would be assumed to follow the same patterns. Obviously the demand for these two goods may be quite different because the first is a good that is sold for consumption while the latter is an intermediate input to a production process. Although, the SIC's are set up to classify production in a number of cases they are often not sufficiently precise to distinguish the differences in demand for various goods.

Estimation

The model estimated is a doubly truncated Tobit regression due to the upper truncation at 1 and the lower truncation at zero of IIT. The Tobit model specification, as defined in Tobin (1958), implies a data generating process defined by a censored normal distribution. The present specification assumes that IIT_{jk} is observed to be non-zero when an unobserved variable y (a propensity, or tendency, to engage in intra-industry trade) is greater than zero and that $IIT_{jk} = 0$ when y < 0 and that y is normally distributed. The implication that y is negative can be interpreted as a negative propensity for a country to engage in intra-industry trade. Thus, the more dissimilar two countries' technologies the more negative the magnitude of the propensity (the less the tendency) to engage in intra-industry trade. Note that we use the Tobit model in this application because when modelling sector level IIT we encounter many country pairs with zeros for IIT_{jk} and the logit models that have been previously used to model IIT do not explicitly model zero values.

In order to compare the estimates of the parameters we have converted the estimated parameters of the Tobit models to the equivalent partial derivatives of IIT with respect to the regressors so that they correspond to the usual interpretation of the parameters in a regression, we will refer to these modified regressors as γ_i (see Nakamura and Nakamura 1983).

Clusters of Parameter Estimates

We estimate over 4,500 parameters (92 parameters for 49 industries) plus over 200,000 variance and covariance values. In order to describe our results we will determine groups of industries that resemble each other with respect to the estimated modified parameters of the model. We employ a unique form of cluster analysis which is designed specifically to handle the problem of combining parameter estimates. This analysis is similar in spirit to the cluster analysis performed by Leamer (1984) to aggregate commodity groups based on the estimated set of regression coefficients from a cross-section model of net exports as a function of country resources.

The cluster method employed here uses an agglomeration or hierarchical strategy for creating the clusters. This means that the individuals start in their own clusters. The first cluster is established from the two closest parameter sets. Once they have been combined the dissimilarities are recomputed and process continues until all individuals are put into the same cluster. We employ a distance measure based on a Chi-square distributed statistic similar to the familiar Wald Test (Wald 1943) used to test parameter differences. This method is related to a much simpler distance measure employed by Nicol (1991) based on a t-statistic. Once membership in the cluster is established the combination of the parameters and their corresponding estimate of the covariance is done by using a mixing formula that has a Bayesian interpretation.

The dissimilarity or "distance" between the parameters estimated for two industries is defined as:

$$D(i,j) = \frac{(\Gamma_i - \Gamma_j)' \, (\Sigma_i + \Sigma_j)^{-1} \, (\Gamma_i - \Gamma_j)}{n} \tag{6}$$

where n is the number of parameters and Σ_i is the estimated covariance of Γ_i. This value is equivalent to the statistic used to test the hypothesis that Γ_j and Γ_i are equal, under the assumption that they are both parameter sets distributed according to independent multivariate normal distributions. $D(i,j)$ is equivalent to a Wald test statistic and is asymptotically distributed as a Chi-square random variable with n degrees of freedom (n equals the number of parameters). It can be seen that high estimated variance for the parameters can offset large differences in the parameters and may result in lower $D(i,j)$ than the value computed for estimated parameters that are closer in value but that are estimated with more precision.

The third element in a cluster analysis is the method for the combination of the parameters into a new cluster. In the present case the cluster location is defined as the parameter values that would have occurred if we combine the data to form a single estimate. Here we use a matrix weighted average of the parameters weighted by their precision matrices (Chamberlain and Leamer 1976). Thus the formula for the new cluster's location is:

$$\Gamma_k = (\Sigma_j^{-1} + \Sigma_i^{-1})^{-1} \, (\Sigma_j^{-1} \Gamma_j + \Sigma_i^{-1} \Gamma_i) \tag{7}$$

where Γ_k is the combination of the ith and jth clusters. The covariance for the kth cluster is given as:

$$\Sigma_k = (\Sigma_j^{-1} + \Sigma_i^{-1})^{-1} \tag{8}$$

Once this new cluster location is formed we assume that because it is a linear function of two other normally distributed random variables, it is normally distributed as well.

The last step in a cluster analysis is the development of a stopping rule to establish the number of clusters to be formed. In this case the change in distances can be defined as:

$$\Delta D_t = D_{t-1} - D_t \tag{9}$$

where D_t is the distance between the two clusters that were combined to create t clusters. Thus, the difference in the cluster distances will be positive because the distances between the remaining clusters become further and further apart. Among the stopping rules suggested by Mojena (1977) is one based on the analogy to a trend in time series analysis. By looking at the first differences we can determine the point where the "trend" shifts.

Regression Results

This section describes the adjusted parameter estimates for the major economic variables in the model. Table 9.1 lists the adjusted parameter values for INEQGDC, GDPSIZE, GDC, DEX, BORDER, EC, and EFTA.

From Table 9.1 we see a wide variety by sector in the regression parameter for INEQGDC. For a number of industries we estimate positive parameter values, indicating a higher level of intra-industry trade between those countries with larger differences in their GDC, thus indicating that the more dissimilar the economy the higher the IIT in: miscellaneous food products (2099a), macaroni (2098), dehydrated vegetables (2034b) and canned fish (2091). It is possible that these signs indicate that these goods are better described by the neoclassical model of trade in goods for which countries have differing resource endowments/environments. In confirmation of the Helpman-Krugman (1985) hypothesis we find that 27 of the 49 sectors have a parameter for INEQGDC that is significantly negative. The greatest impact is estimated for beet sugar (2063), dry milk (2023), sausages (2013a) and butter (2021).

From the Helpman-Krugman (1985) model we expect the estimated parameters on GDPSIZE to be positive. However, from Table 9.1 we note that the majority of these parameters are not significantly different from zero at a 5 percent level of significance. Of the five that are significantly different from zero only pickles, sauces, and salad dressing (2035) are negative.

Again we see that IIT for each commodity will be effected in a different way by the GDC of the exporting country. Note that under the Helpman and Krugman (1985) hypothesis we expect that these parameters to be positive. However, of the 20 parameters that are estimated to be significantly different from zero at the 5 percent level, 9 are negative. The primary sectors that are negative are: bottled soft drinks (2086), cheese (2022), sugar and candy (2060), and miscellaneous food preparations (2099a) all report significantly negative coefficients for these parameters. Among the sectors with positive values are beet sugar

TABLE 9.1 The Estimated Adjusted Coefficients

Sector	INEQGDC	GDPSIZE	GDC	DEX	BORDER	EC	EFTA
2011a, Meat Packing	-0.107*	-0.361	-0.603**	-0.013	0.183**	0.075**	0.075**
2011b, Hides, Skins & Furs	-0.480**	0.631	0.214	-0.040*	0.250**	-0.057**	0.084**
2013a, Sausages & Prep Meats (Smoked)	-0.889**	1.687	-0.343	0.011	0.187**	0.126**	0.074**
2013b, Sausages & Prep Meats (Tinned)	-0.362**	0.720	-0.401	-0.024	0.196**	0.078**	0.031**
2015, Poultry	-0.324**	-2.647	0.358	-0.024	0.197**	0.224**	0.040*
2021, Creamery Butter	-0.692**	-5.450	0.932*	-0.047	0.184''	0.241''	0.037
2022, Cheese	-0.156*	0.062	-1.143**	-0.058**	0.125**	0.049**	0.052**
2023, Dry & Condensed Milk	-1.003**	0.380	-0.224	-0.050*	0.247**	0.127**	0.125**
2026, Fluid Milk	-0.433**	26.464	0.559	-0.170*	0.347**	0.268**	0.000
2032, Canned Soups	-0.101	7.056*	-0.532	-0.035	0.200**	0.038**	0.084**
2033a, Canned Fruits	-0.088*	1.841*	-0.782**	0.015	0.208**	0.053**	0.154**
2033b, Canned Vegetables	-0.018	0.112	-0.033	-0.020	0.196**	0.026**	0.184**
2034a, Dehydrated Fruits	-0.039	-0.298	0.696**	-0.018	0.216**	0.080**	0.105**
2034b, Dehydrated Vegetables	0.161**	0.560	0.624**	0.005	0.133**	0.081**	0.150**
2035, Pickles & Sauces	0.001	-2.058*	0.778**	-0.060**	0.191**	0.116**	0.229**
2037a, Frozen Fruits	-0.234	3.988	-0.169	-0.019	0.282**	0.092**	0.080**
2037b, Frozen Vegetables	0.011	2.683	0.692*	-0.086**	0.215**	0.115**	0.086**
2041, Flour & Grain Mill Prod	-0.122**	0.010	1.130**	-0.022	0.220**	0.082**	0.089**
2043, Cereal Breakfast Foods	-0.071	-0.463	0.245	-0.024	0.234**	0.055**	0.108**
2044, Rice Milling	-0.062	9.635*	0.161	-0.021	0.180**	0.122**	0.101**
2046, Wet Corn Milling	-0.328**	36.447	-0.223	-0.179**	0.294**	0.121**	-0.148**
2048, Prepared Feeds, NEC	-0.620**	1.028	-0.350	-0.049**	0.211**	0.069**	0.080**
2051a, Bread & Biscuits	-0.385**	-0.025	-0.613	0.007	0.302**	0.061**	0.108**
2051b, Cake & Pastry	-0.154*	-1.565	0.164	-0.044*	0.229**	0.077**	0.209**

Note: * Significantly nonzero at the 5 percent level.
** Significantly nonzero at the 1 percent level.

TABLE 9.1 (continued)

Sector	INEQGDC	GDPSIZE	GDC	DEX	BORDER	EC	EFTA
2060, Sugar, Candy & Confec Prod	-0.181**	-0.876	-0.863**	-0.036**	0.167**	0.093**	0.113**
2063, Beet Sugar	-1.372*	23.288	1.566	-0.112	0.225**	0.232**	-0.054
2066a, Cocoa Products	0.057	1.973	0.425	-0.008	0.182**	0.069**	0.134**
2066b, Chocolate	-0.366**	-0.369	-0.630**	-0.051**	0.255**	0.119**	0.186**
2068, Salted & Roasted Nuts, Seeds	-0.104**	0.265	0.376*	-0.016	0.140**	0.054**	0.175**
2074, Cottonseed Oil Mills	-0.225	2.363	1.337*	0.077	0.116**	0.060**	-0.394
2075, Soybean Oil Mills	-0.194**	9.552*	0.426*	-0.028	0.217**	0.002	0.113**
2076, Vegetable Oil Mills, NEC	-0.062	0.120	0.003	-0.013	0.180**	0.076**	0.098**
2077a, Meat or Fish Meal Fodder	-0.204**	1.656	-0.593**	-0.194**	0.227**	0.064**	0.056**
2077b, Animal & Marine Fats & Oils	-0.125**	0.086	0.563*	0.024	0.252**	0.044**	0.089**
2079a, Margarine, Shortening	-0.528**	-1.105	0.824**	-0.115**	0.227**	0.092**	0.074**
2079b, Fixed Vegetable Oil	-0.179**	0.552	0.147	0.007	0.228**	0.070**	0.135**
2079c, Processed Animal & Veg Oil	-0.144**	0.878	-0.138	-0.005	0.211**	0.008	0.171**
2082, Malt Beverages	-0.040	-0.005	-0.288	0.003	0.171**	0.031**	0.099**
2083, Malt	-0.430	50.754	1.477	-0.005	0.273**	0.133**	0.136**
2084, Wines & Brandy	-0.126**	0.281	0.257	-0.025**	0.166**	-0.035**	0.056**
2085, Distilled & Blended Liquors	-0.009	-0.686	-0.080	-0.028**	0.141**	-0.004	0.081**
2086, Bottled & Canned Soft Drinks	-0.027	2.076	-1.250**	-0.066**	0.318**	0.078**	0.164**
2091, Canned & Cured Seafoods	0.157**	1.131	-0.207	-0.034**	0.099**	0.064**	0.073**
2092, Fresh or Frozen Fish	0.078*	-0.624	-0.341	0.000	0.209**	0.046**	0.090**
2095, Roasted Coffee	0.056	0.461	-0.548*	-0.081**	0.161**	0.036**	0.099**
2098, Macaroni & Spaghetti	0.395**	6.665	-0.202	-0.050**	0.264**	0.102**	0.091**
2099a, Misc Food Preparations	0.419**	5.286*	-0.783**	-0.008	0.223**	0.041**	0.140**
2099b, Tea & Mate	0.052	3.587	0.236	-0.024	0.248**	0.098**	0.125**
2099c, Spices	-0.146**	0.816	0.647**	0.002	0.120**	0.099**	0.187**

Note: * Significantly nonzero at the 5 percent level.
      ** Significantly nonzero at the 1 percent level.

(2063), cottonseed oil (2074), flour (2041) and butter (2021). The high proportion of insignificant and contrary results for this coefficient may be due to potential collinearity with the dummy variables for the exporting countries. Again we have a difficult time establishing which commodities are similar in their confirmation to the hypotheses concerning IIT.

Meat or fish meal fodder (2077a) appears to be the most sensitive to exchange rate volatility (DEX). Only animal and marine fats and oils (2077b) appear to counter our preconceived notion of the negative effect of exchange rate volatility. This large variation for these two related sectors demonstrates the importance of the segmentation of the SIC=2077 correspondence to the SITC. It is also possible that the perverse signs on a few products reflect the specific set of countries that trade in these sectors.

Cluster Results

The parameters in Table 9.1 have varying degrees of confidence attached to the parameter estimate as given by the t-statistics. Thus, even if we only use the parameter estimates from this one regressor to form conclusions concerning the similarity of various commodities, it will be necessary to establish a procedure for incorporating the precision of the estimate as well as the parameter estimate. The conflict as to how to incorporate the t-statistics and the difficulty in the interpretation of a set of up to 92 columns or even a subset of the parameters, is precisely the problem that the cluster method is designed to solve. In the following analysis we cluster the industries into groups based on the parameters with specific economic interpretation. We use the estimates and the corresponding estimated covariance for the values of γ_1 to γ_{10} (the adjusted tobit regression coefficient values that correspond to the partial derivatives with respect to the first 10 regressors). Then we report the values for these parameters for each cluster of commodities we define. First we determine the number of clusters to be used, then we describe the membership in each cluster, and last we report the aggregate parameter values for each cluster.

By plotting the differences in the distances (ΔD_t) as we aggregate up the number of clusters we found a point where a large shift occurs. We found that with 10 to 9 clusters we get a major shift in the first differences of the distances. Thus, we will report the membership in 10 clusters.

Table 9.2 reports the cluster membership for each of the 10 clusters. Note that a number of clusters have only one member. Clusters 2, 7, and

8 are defined by only one commodity. Clusters 1, 3, 4, and 6 include the majority of the rest of the industries.

To see how the clusters are formed we provide the icicle plot for the clusters in Figure 9.1 (a discussion of the interpretation of these plots can be found in Kruskal and Landwehr 1983). The number of clusters is on the vertical axis and the cluster membership is on the horizontal axis. This figure shows how the clustering method combines the clusters from the case where each is in its own cluster (when there are 49 clusters at the far left-hand edge of the figure) to the case in which all industries are in the same cluster (at the top). For example, from this diagram we can see that the first two industries to be combined into a single cluster (the move from 49 to 48 clusters) were spices (2099c) and cottonseed oil (2074). The transition from 48 clusters to 47 clusters was accomplished by the inclusion of 2068 salted and roasted nuts and seeds into the cluster formed by 2099c and 2074. This diagram can also be used to investigate the genealogy of the 10 clusters defined in Table 9.2. By going across the figure on a horizontal line from the 10 cluster point we see that the first two industries 2098 and 2099a in cluster 10. This diagram shows that if we had stopped the clustering at 18 clusters each of these industries would be in clusters with only one sector. This diagram also points out a potential problem with agglomeration clustering - that once the clustering algorithm puts an individual in a cluster it will never be removed and put into another.

Table 9.3 lists the parameter estimates and the corresponding t-statistics for the industries included in each cluster and the parameters for a cluster of the whole set (the case where there is only one cluster). The parameters and the covariance have been computed using the matrix weighted average described above. From Table 9.3 we can see that cluster 7 and 10 contradict the Helpman-Krugman (1985) hypothesis that predicts a negative coefficient on INEQGDC. These are not major industries within the SIC=20 group and, again, the trade in these industries may best be described by neoclassical trade theory. In addition, the catch-all characteristic of the 2099a sector may be responsible for the nonconformity of these estimates to the Helpman-Krugman (1985) hypothesis, as well.

We find a negative influence of membership in the European Community (EC) for cluster 2 which is the hides and skins portion of the 2011 sector, confirming our assumption that this sector should be separated from the meat packing part of the 2011 SIC to perform an analysis of the factors that affect foreign trade. We also find a negative coefficient for EFTA in cluster 5 (which is composed of the fluid milk and

FIGURE 9.1 Icicle Plot of Clusters by Number.

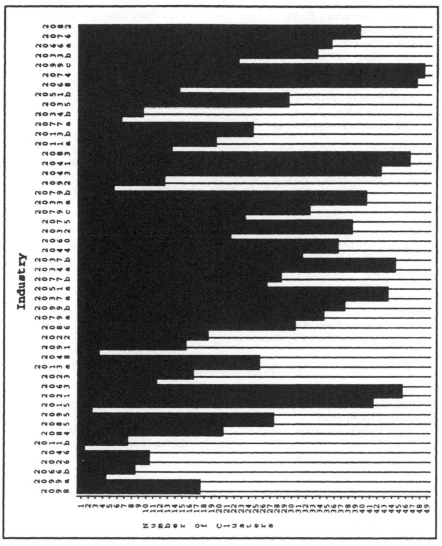

TABLE 9.2 Cluster Membership

CLUSTER=1

2011a, Meat Packing Plants
2013b, Sausages & other Prepared Meats (Tinned)
2041, Flour & other Grain Mill Products
2043, Cereal Breakfast Food
2077a, Meat or Fish Meal Fodder
2083, Malt Extract
2092, Fresh or Frozen Packaged Fish

CLUSTER=2

2011b, Hides, Skins & Furs Undressed

CLUSTER=3

2013a, Sausages & other Prepared Meats (Smoked)
2015, Poultry Slaughtering & Processing
2021, Butter
2023, Dry, Condensed, Evaporated Products
2048, Prepared Feeds, NEC
2063, Sugar Beet Fresh Dry, Cane

CLUSTER=4

2022, Cheese & Curd
2032, Canned Soups
2033a, Canned Fruits
2034a, Dehydrated Fruits
2037a, Frozen Fruits
2037b, Frozen Vegetables
2044, Rice Milling
2051a, Bread & Biscuits
2060, Sugar, Candy & Confectionery Products
2075, Soybean Oil Mills
2077b, Animal & Marine Fat & Oil
2079a, Margarine, Shortening
2079b, Fixed Vegetable Oil
2079c, Processed Animal & Veg Oil
2086, Bottled & Canned Soft Drinks
2091, Canned & Cured Fish & Seafood
2099b, Tea & Mate

CLUSTER=5

2026, Fluid Milk
2046, Wet Corn Milling

CLUSTER=6

2033b, Canned Vegetables
2035, Pickles, Sauces & Salad Dressing
2051b, Cake & Pastry
2066a, Cocoa Products
2068, Salted & Roasted Nuts & Seeds
2074, Cottonseed Oil Mills
2076, Vegetable Oil Mills, NEC
2082, Malt Beverages
2099c, Spices

CLUSTER=7

2034b, Dehydrated Vegetables

CLUSTER=8

2066b, Chocolate

CLUSTER=9

2084, Wines, Brandy & Brandy Spirits
2085, Distilled Liquors
2095, Roasted Coffee

CLUSTER=10

2098, Macaroni & Spaghetti
2099a, Misc Food Preparations

TABLE 9.3 Parameter Estimates and t-statistics by Cluster

CLUSTER	INEQGDC	GDPSIZE	GDC	DEX	BORDER	EC	EFTA
1	-.17594	.1140	-.06342	-.02609	.20797	.07404	.06991
	-8.6391	.30701	-.67772	-3.85357	64.695	2.4426	15.7902
2	-.47981	.6311	.21433	-.03961	.24998	-.05682	.08450
	-11.2806	.82076	.99352	-2.37524	3.7260	-6.2129	8.5471
3	-.74366	.8610	-.03614	-.02977	.20423	.14140	.08030
	-22.1135	.70861	-.26549	-3.04066	52.027	29.3669	12.9519
4	-.09967	.3770	-.11805	-.01674	.20852	.06620	.10355
	-8.0418	1.54557	-2.05489	-4.73467	101.398	28.8855	37.9411
5	-.36195	29.7251	.08559	-.18487	.30774	.16984	-.11031
	-3.8539	1.94330	.25130	-3.66181	35.790	15.2216	-6.1740
6	-.07007	-.1943	.26047	-.01469	.17114	.06436	.15937
	-4.5162	-.69476	3.49654	-3.42030	61.070	2.5862	41.5911
7	.16110	.5601	.62445	.00479	.13272	.08064	.14971
	4.1874	.49699	3.04579	.45940	14.8315	8.5829	13.1809
8	-.36645	-.3695	-.63015	-.05143	.25494	.11872	.18624
	-6.4679	-.57305	-2.74403	-4.19791	3.6011	12.9995	18.8064
9	-.03089	.0576	-.06862	-.03910	.15842	-.00602	.07573
	-1.3234	.20634	-.62725	-6.54827	37.096	-1.2840	13.4065
10	.40212	5.6559	-.54214	-.01887	.24483	.06982	.12089
	12.0786	2.88917	-3.27736	-1.93386	37.062	9.7271	13.5549
All	-.10314	.1583	-.1878	-.01975	.19972	.06598	.10597
	-14.2849	1.21726	-.55479	-9.53751	163.990	48.2369	64.0604

Note: t-statistics are reported below coefficient estimates for each cluster.

wet corn milling industries). In looking at the estimates for this parameter from both regressions we find that the wet corn milling sector has the only significantly negative parameter for the EFTA dummy. The fluid milk sector is included because the estimated parameter for mutual membership in EFTA is estimated as .0002 with a t-statistic of .0058.

Cluster 4 includes the largest collection of industries. This cluster is characterized by a significant and negative coefficient on INEQGDC an insignificant coefficient for GDC and a significantly negative coefficient on the importing country's income per capita (GDC). This last result may be due to the potential for the country dummies to swamp the impact of the parameter. There is no obvious set of industries in this cluster that would have more of an environmental advantage than any other cluster. Cluster 6 appears to contain that group of industries with parameter values that best coincide with those predicted by the Helpman-Krugman (1985) hypothesis. Except for spices (2099c), and salted and roasted nuts (2068), this cluster appears to contain many industries that do not exhibit any environmentally based advantages. And if the majority of the value traded in the spices sector is made up of repackaged raw commodities that are imported from less developed countries, all the industries in cluster 6 demonstrate the characteristics of industries that are consistent with the Helpman-Krugman (1985) model. Cluster 3, although it does not result in a sign for the parameter on GDC that is significantly positive, is estimated with the largest impact of the INEQGDC variable. Compared to cluster 6, the inequality variable has over ten times the impact on IIT. This cluster contains a number of industries that can be located in any of the countries in the sample. Thus, allocation of the sector in this cluster is independent of environmental factors.

The last line in Table 9.3 is a set of parameters for a cluster containing the entire set of SIC=20 industries. It is equivalent to the analysis done in HSD, although these parameters are the adjusted values for γ and not the estimated parameters β. We can compare the signs. We find that the aggregate results do not reject the Helpman-Krugman (1985) hypothesis. However, the parameters on both GDPSIZE and GDC are not significantly different from zero. In the HSD paper the GDC parameter was estimated as significant and positive. The large t-statistics on many of the parameters listed in the "All" row of Table 9.3 is due to the large number of observations. These parameters are equivalent to the parameters that we would obtain by fitting a model with common parameters for the first 10 regressors then estimating separate intercepts, variances, exporting country and importing country dummies and separate year dummies for each sector.

In sum, we find that 10 sectors in clusters 2, 5, 7, 9 and 10 can be singled out as having differing characteristics of their intra-industry trade than the remaining 39 sectors that make up the other 4 clusters.

Summary and Conclusions

In this chapter we have modelled the level of intra-industry trade in various sectors of the processed food industry. Although the original specification of the Helpman-Krugman (1985) model was formulated using examples from durable goods manufacturing industries, it has become increasingly obvious that sectors of the economy that produce food products also engage in IIT. The results from this work can be used to determine the extent to which one can observe IIT among different sectors in the SIC=20 industry. We find that certain industries are significantly more prone to engage in IIT given the increasing similarity of the technology of the countries that trade than others, thus implying that policies that affect trade may be more specifically directed to industries that will have the greatest potential for growth in this area.

Further research in this area will focus on the possible clusters formed by other parameters estimated in these models. For example, which industries are similar in the time factors and how do they differ. Or we could cluster the exporting and importing country dummies to establish how the model is picking up any country specific effects caused by trade policies or other taste/production conditions in one country or another. Other future research may involve the reestimation of the basic model with different SIC/SITC aggregations and the consideration of cross country models where we fit models for each country across all industries and cluster countries.

Notes

1. The modified SIC definitions used are available from the authors. Sectors are also indicated in Table 9.2.

References

Balassa, B., and L. Bauwens. 1987. "Intra-Industry Specialization in a Multi-Country and Multi-Industry Framework." *Economic Journal* 97:923-939.

Bergstrand, J. H. 1990. "The Heckscher-Ohlin-Samuelson Model, The Linder Hypothesis, and the Determinants of Bilateral Intra-Industry Trade." *Economic Journal* 100:1216-1229.

Chamberlain, G. and E. E. Leamer. 1976. "Matrix Weighted Averages and Posterior Bounds." *Journal of the Statistical Society* B. 38:73-84.

Christodoulou, M. 1992. "Intra-industry Trade in Agrofood Sectors: The Case of the EEC Meat Market." *Applied Economics* 24:874-884.

Dayton, J. R., and D. R. Henderson. 1992. "Patterns of World Trade in Processed Food." Unpublished Working Paper, The Ohio State University.

De Grauwe, P., and B. de Bellefroid. 1987. "Long-Run Exchange Rate Variability and International Trade," in S. W. Arndt and J. D. Richardson, eds., *Real-Financial Linkages among Open Economies*. Cambridge, MA: MIT Press.

Dixit, A.K., and V. Norman. 1980. *Theory of International Trade*. Cambridge: Cambridge University Press.

Greenaway, D. and Milner, C.R. 1986. *The Economics of Intra-Industry Trade.* Oxford: Blackwell.

Grubel, H.G. and P.J. Lloyd. 1975. *Intra-Industry Trade.* London: Macmillan.

Hart, T., and B. J. McDonald. 1992. *Intra-Industry Trade Indexes for Canada, Mexico and the United States, 1962-87,* Staff Report No. AGES 9206, Agriculture and Trade Analysis Division, Economic Research Service, U.S. Department of Agriculture.

Havrylyshyn, O., and E. Civan. 1984. "Intra-Industry Trade and the State of Development," in P. K. M. Tharakan, ed., *The Economics of Intra-Industry Trade.* Amsterdam: North Holland.

Helpman, E. 1981. "International Trade in the Presence of Product Differentiation, Economics of Scale and Monopolistic Competition." *Journal of International Economics* 11:305-340.

_____. 1987. "Imperfect Competition and International Trade: Evidence from Fourteen Industrial Countries." *Journal of the Japanese and International Economies* 1:62-81.

_____, and P. R. Krugman. 1985. *Market Structure and Foreign Trade: Increasing Returns, Imperfect Competition and the International Economy.* Cambridge, MA: MIT Press.

Hirschberg, J. G., I. M. Sheldon and J. R. Dayton. 1994. "An Analysis of Bilateral Intra-Industry Trade in the Food Processing Sector." *Applied Economics* 26:159-167.

Krugman, P. R. 1979. "Increasing Returns, Monopolistic Competition, and International Trade." *Journal of International Economics* 9:469-479.

_____. 1981. "Intra-Industry Specialization and the Gains from Trade." *Journal of Political Economy* 95:959-973.

Kruskal, J. B. and J. M. Landwehr. 1983. "Icicle Plots: Better Displays for Hierarchical Clustering." *The American Statistician* 37:162-168.

Leamer, E. E. 1984. *Sources of International Comparative Advantage.* Cambridge, MA: The MIT Press.

Linder, B. 1961. *An Essay on Trade and Transportation.* New York, NY: John Wiley.

McCorriston, S., and I. M. Sheldon. 1991. "Intra-Industry Trade and Specialization in Processed Agricultural Products: The Case of the US and the EC." *Review of Agricultural Economics* 13:173-184.

Mojena, R. 1977. "Hierarchical Grouping Methods and Stopping Rules: An Evaluation." *Computer Journal* 20:359-363.

Nakamura, A. and M. Nakamura. 1983. "Part-time and Full-time Work Behavior of Married Women: A Model with Doubly Truncated Dependent Variable." *Canadian Journal of Economics* 16:229-257.

Nicol, C. J. 1991. "The Effect of Expenditure Aggregation on Hypothesis Tests in Consumer Demand Systems." *International Economic Review* 32:405-416.

Noland, M. 1988. "Econometric Estimation of International Intraindustry Trade." Unpublished Working Paper, Institute for International Economics.

SAS Institute Inc. 1989. *SAS/STAT® User's Guide, Version 6,* Fourth Edition, Volume 2. Cary, NC: SAS Institute Inc.

Summers, R. and A. Heston, 1991 "The Penn World Table (Mark 5): An Expanded Set of International Comparisons, 1950-1988." *Quarterly Journal of Economics* 106:327-368.

Tharakan, P. K. M. 1985. "Empirical Analyses of the Commodity Composition of Trade," in D. Greenaway, ed., *Current Issues in International Trade: Theory and Policy.* London: Macmillan.

Tobin, J. 1958. "Estimation of Relationships for Limited Dependent Variables." *Econometrica* 26:24-36.

Wald, A. 1943. "Tests of Statistical Hypotheses Concerning Several Parameters When the Number of Observations is Large." *Transactions of the American Mathematical Society* 54:426-82.

Wickham, E. and H. Thompson. 1989. "An Empirical Analysis of Intra-Industry Trade and Multinational Firms," in P.K.M. Tharakan and J. Kol, eds., *Intra-Industry Tade: Theory, Evidence and Extensions.* New York, NY: St. Martins Press.

10

The Determinants of Intra-Firm International Trade

Kai Wang and John M. Connor

Introduction

Intra-firm trade (IFT) by multinational enterprises (MNEs) was first mentioned by Grubel and Lloyd (1975). Empirical studies appeared a few years later (Lall 1978; Helleiner and Lavergne 1979).[1] IFT is the international transfer of products or services between business units under common ownership. By definition, IFT, also called "related-party trade," refers to trade between parent companies and their *foreign* affiliates or between affiliates of the same parent company.

Among home countries of MNEs, only the United States regularly publishes extensive data on MNE-related trade, of which IFT is a substantial portion.[2] In 1957, IFT accounted for about 20 percent of total U.S. merchandise imports and exports. Increasing steadily, IFT accounted for about 30 percent of total U.S. exports and 40 percent of imports by the mid 1980s (United Nations 1988). U.S. manufacturing MNEs have strong preferences for IFT. From 46 to 64 percent of U.S. parents' trade is intra-firm (Table 10.1). For U.S. manufacturing affiliates abroad, IFT approaches 90 percent of total trade; for joint ventures abroad, IFT is lower.[3] IFT by U.S. food manufacturers is much lower -- about half that for other manufacturers. Table 10.1 also measures IFT between U.S.-based, foreign-owned companies and their non-U.S. parents. For these U.S. affiliates, the majority of U.S. imports are purchased from the overseas parent group, whereas only a minor portion of U.S. affiliate exports goes to the parent group.[4] U.S. affiliate exports is the one instance where IFT by food processors is greater than that of all manufacturers (39 percent versus 29 percent).

TABLE 10.1 Alternative Measures of the Degree of Intra-Firm Trade, by Multinational Manufacturing Companies and Food Manufacturing Companies, 1987 and 1989

	Type of Trade			Exports		Imports	
No.	Value of Intra-Firm Trade (Numerator)	Value of Total Trade (Denominator)	Year	Mfg.	Food Mfg.	Mfg.	Food Mfg.
				Percent			
1.*	U.S. parents with all their foreign affiliates	U.S. parents' U.S. trade	1989	45.6	15.1	55.4	26.8
2.**	U.S. parents with all their foreign affiliates	U.S. parents' U.S. trade	1989	48.7	21.4	63.8	33.6
3.**	U.S. parents with their majority-owned foreign affiliates	Majority-owned foreign affiliates' U.S. trade	1989	89.1	72.3	87.5	69.6
4.**	U.S. parents with their minority-owned foreign affiliates	Minority-owned foreign affiliates' U.S. trade	1989	76.3	57.6	45.8	41.7
5.***	All U.S. affiliates with non-U.S. parent group	Total U.S. trade by U.S. affiliates, by affiliates' industry	1987	29.0	38.7	63.4****	55.3****
6.***	All U.S. affiliates with non-U.S. parent group	Total U.S. trade by U.S. affiliates, by product	1987	39.7	37.8	79.1	41.2

Notes: Unless stated otherwise, manufactures, manufactures and food manufactures are all of the products and services traded by enterprises primarily classified in the respective industry. Definition 6 differs from this general rule. Parents and affiliates are nonbank companies.

* These data are as reported by the U.S. parents themselves.

** These data are as reported by the foreign affiliates of U.S. parents. The affiliates report slightly higher U.S. exports from their parents (0.1% higher) and U.S. imports to their parents (3.4% higher) than their parents do for three reasons: differences in timing, in valuation, and whether certain small affiliates are exempt from reporting themselves but are not exempt when the parents report.

*** These data reported by U.S. affiliates.

**** U.S. imports to U.S. affiliates are of two types: final products for resale (23% of value shipped by manufacturers and 38% by food manufacturers) and intermediate products (77% and 62%). Intra-firm U.S. imports are higher for final foods (60%) than for intermediate food materials (52%). For all manufactures, there is no difference (BEA 1987:Table C-35).

Sources: For 1-4, (BEA 1989:Tables II.Q1 and II.Q4); for 5 (BEA 1989:Table III.F3); and for 6 (BEA 1989:Table III.F14).

IFT is more intensive for trade with developed market economies than for trade with developing or previously centrally planned economies (Helleiner 1979). IFT is more intense in high technology industries such as chemicals, machinery, and transport equipment. IFT with developed economies typically involves sales of final products, though a sizeable proportion of IFT is in parts and components. However, IFT from the developing countries is largely in raw materials.

IFT is one way MNEs integrate economic activities across geographical, cultural, and legal boundaries to achieve global profit maximization. IFT may be seen as the replacement of trade through markets by unrelated agents with trade within a command system. In traditional theory, trade is assumed to take place at "arm's length" between unrelated buyers and sellers in a perfectly competitive market. In the case of IFT, the market mechanism is replaced by vertical integration through ownership. While most economists presume IFT behaves differently from arm's length trade, Benvignati (1990) found factors expected to explain IFT did not perform differently for intra-firm exports (IFX) than for overall exports. Thus, questions remain as to whether IFT is different from arm's length trade and what accounts for the differences.

The major policy interest in IFT is its relationship to MNE business strategies, and to transfer pricing in particular. Transfer pricing is the practice of geographically reallocating profits by over- or under-invoicing the stated values of IFT to avoid tax, tariffs, and other fiscal regulations in one or more legal jurisdictions. A 1990 U.S. House Ways and Means Committee report charged that the U.S. Treasury was losing $30 billion per year because of transfer pricing by U.S. subsidiaries of foreign multinationals (Bucks 1991). Japanese-owned affiliates, among the highest in IFT, are prime suspects (Lawrence 1991).

The purpose of this paper is to test the relative predictive accuracy of three mutually inconsistent theories of intra-firm trade. The three paradigms are transactions-cost, industrial-organization, and transfer-pricing approaches. We examine determinants of both IFX and intra-firm imports (IFM) using U.S. data on manufacturing industries from the 1980s, paying particular attention to food product industries.

Theory

Trade theory in the Hecksher-Ohlin tradition fails to explain certain crucial features of IFT as a consequence of international production by MNEs. There are many conceptual parallels between IFT and FDI. Just as IFT replaces international trade in goods through markets with trade by a command system, so too FDI replaces international transfers of financial assets through bond markets with transfers of equity capital.

Initial attempts to explain FDI in general made few distinctions between direct and portfolio international capital flows. These were soon abandoned due to the failure to explain the transfer of resources (e.g., technology and management) other than financial capital internally within the firms rather than externally between independent parties. The earlier theories were also unable to explain geographic cross-penetration, geographic concentration, and industrial patterns of FDI flows (Connor 1977). What emerged was a wholly different approach to firm and industrial organization that focussed on the motives for global expansion. The starting point is the following proposition: compared to local competitors, MNEs entering a market from abroad are faced with additional decision-making costs arising from cultural, legal, and linguistic differences; lack of knowledge of local market conditions; communication expenses; and other information-based disadvantages. In order for MNEs to be profitable, there must be some firm-specific advantages to offset these extra costs. These advantages should be transferable at very low marginal costs within firms and across geographic market boundaries (Caves 1971).

Hymer (1960) gave primacy to market structure in explaining FDI. In Hymer's approach, FDI is a form of oligopolistic conduct arising from imperfectly structured markets. A necessary condition for FDI is that MNEs acquire and sustain firm-specific monopolistic advantages in home markets. As later elaborated by Kindleberger (1969) and Dunning (1979), these ownership-specific advantages of firms must be linked to the locational advantages of home and host countries. Locational factors include trade barriers, host government policies, relative labor costs, market size, and relative demand growth. The need for MNEs to generate firm-specific advantages through R&D was emphasized by Vernon (1966). Vernon hypothesized that, since the firm-specific advantage of MNEs is in constant danger of being eroded, research and development (R&D) was necessary to overcome entropy to generate new advantages. Many studies subsequently found that R&D intensity was an important determinant of FDI flows, but the life-cycle theory itself has proven difficult to verify (Connor 1983).

Knickerbocker (1973) demonstrated significant temporal clustering of initial moves abroad by U.S. MNEs, attributed to oligopolistic reactions. Using a framework based on a repeated, noncooperative game, Graham (1990) conceptualized intra-industry FDI as an "exchange-of-threat" phenomenon of rivalry among MNEs. Graham's model is similar to the reciprocal dumping model developed by Brander and Krugman (1983).

Knickerbocker's (1973) follow-the-leader pattern of FDI has been confirmed by other empirical studies.

Coase (1937) recognized that market transactions incurred costs, such as finding a relevant price, contract negotiations, agent fees, and taxes. Under some conditions it is more efficient for firms to undertake transactions within the firm rather than through established market mechanisms. Williamson (1975, 1989) provides a modern treatment of Coase's ideas.

These partial explanations of FDI were synthesized in Dunning's (1979) "eclectic" OLI model (Ownership-Location-Internalization advantages). In pursuing a systemic explanation of MNEs activities in terms of the firm's ability to internalize international market transactions, Dunning emphasized that firms' ownership advantages and locational aspects can explain selling abroad generally, but these sales might take the form of controlled FDI, licensing, joint ventures, or other looser alliances. For a firm to prefer internalizing foreign transactions, sufficient incentives should include avoidance of transactions costs, negotiations on enforcing property rights, buyer's uncertainty, quality control, and government interventions. Rugman (1980), Sugden (1983), and Ethier (1986) provide increasingly rigorous models of internalization.

To paraphrase Ethier (1986), IFT is likely to occur when the information required to export (1) is too complex or too extensive for an efficient transaction to occur outside the MNE, or (2) has public-good characteristics that frustrate the ability of an MNE to appropriate the revenues necessary to cover costs of production and international marketing. In the second case, the inadequacy of patents or corporate secrecy to protect a proprietary advantage over a new technology is a prime example. Casson (1987) provides a more specific list of the conditions for MNE internalization. They include: the inability of arm's length contracts to cope with rigidities and irreversibilities in the production process; the tendency to distrust substitution decisions concerning those parts of the process that are flexible; monopolistic price discrimination and various entry barriers; the novelty of the division of labor; and incentives for transfer pricing, such as statutory intervention in product markets through price regulation.

Previous Empirical Studies

While many theorists in the fields of trade, finance, management, and industrial organization have attempted to explain IFT as one feature of FDI, difficulties validating various hypotheses have generated a limited empirical literature. The first was a simple cross-sectional regression by

Lall (1978). Using highly aggregated industry-level data provided by the U.S. Tariff Commission, Lall estimated industry differences of IFX by U.S. manufacturing firms in 1970, relying on general concepts of vertical integration. He found that the share of IFX in U.S. MNEs' exports was positively associated with variables representing technological intensity, extent of foreign direct investment, divisibility of production processes, and the need for after-sales service. Lall suggested that estimating the determinants of IFX would be more successful at the product level rather than at the broad industry level. Moreover, Lall was limited to the examination of one year's data and was not able to include other factors he thought might be important, such as risk, uncertainty, scale economies, capacity utilization, and host government policy.

Helleiner and Lavergne (1979) examined IFM by U.S. manufacturers, based on 1975-1977 data on 50 to 100 manufacturing categories. In their econometric test, IFM was positively correlated with R&D expenditures as a proportion of sales, average wage, and firm size. The theoretical underpinnings of their model are obscure. With data disaggregated by area of origin (OECD or Third World), they unsuccessfully tested for the influence of market structure, such as scale economies, product differentiation, and barriers to entry. Helleiner (1979) used a similar model for 1975 U.S. IFM, but his three market-structure variables were not significant. His model explained only about 20 percent of the variation in IFM. Helleiner and Lavergne noted the limitations of assessing the differences between IFT and arm's length trade without introducing more systemic and detailed data.

In a third study, Siddarthan and Kumar (1990) analyzed both IFX and IFM using a model derived primarily from the transactions-cost approach. Using the U.S. Department of Commerce's 1982 Benchmark Survey Data on U.S. Direct Investment Abroad, they examined the determinants of industry variation in the proportion of IFT across 32 branches of manufacturing. They assert that IFX is mostly finished goods, whereas intermediate products dominate IFM. R&D and wages had the same positive impact on IFT that was found in the two previous studies. In addition, Siddharthan and Kumar (1990) found that the extent of parent-company international orientation was positively related to IFX and that anti-pollution regulations in the U.S. positively influence IFM. They fail to distinguish between internalization incentives due to technological reasons and those due to market failure.

Benvignati (1990) had access to the U.S. Federal Trade Commission's Line-of-Business survey. These data have been the basis of most advanced empirical studies in industrial organization in manufacturing published in the 1980s. Data on the value of IFX were collected from 249 leading U.S. firms for the highly inflationary years 1975, 1976, and 1977 and were organized into 498 lines of business (roughly akin to broadly defined company profit centers). Benvignati's model is built primarily on the transactions-cost approach. This study is of limited interest, however, because the principal purpose was to test whether transactions costs explained differences between *total* U.S. export values and *total* IFX by a limited number of very large firms. Benvignati concludes that the explanatory factors behind exports and IFX are the same, but this conclusion is based on a suspect dummy-variable technique.

The Determinants of IFT

In this section we distinguish among three theoretical approaches to understanding the choice an MNE faces between IFT and unrelated-party (conventional) trade. The first framework comes from the industrial-organization tradition and is best represented by the model of Sugden (1983). Although Sugden (1983) assumes a peculiar kind of product homogeneity in deriving his model, we extend his reasoning to the heterogeneous-products case in an informal manner. The second approach is based on Coasian internalization of transactions costs. We rely on the interpretations given by Siddarthan and Kumar (1990) and Benvignati (1990). The third explanation of IFT is based solely on the policy response of MNEs as recipients of rents or quasi-rents from any of several sources. The transfer-pricing school is grounded in the political-economy writings of economists concerned about developing economies (Vaitsos 1974). According to this view, a threshold level of IFT is required for MNEs to maximize *global after-tax* profits. Simply put, IFT is a mechanism for the transfer of reported profits from high-corporate-tax countries to low-tax countries (see also Henneberry and Russell 1989). Tax avoidance and possibly tax evasion is fostered by high effective corporate-tax rates in the U.S. relative to the rest of the world. Thus, whatever factor yields rents will be positively related to the extent of IFT. Although the predictions from oligopolistic conduct, internalization, and transfer-pricing overlap in some cases, they are sufficiently different to provide fairly clear-cut empirical assessment of the paradigms.

Figure 10.1 lists specific hypotheses derived from the transfer-pricing scenario. R&D, K/Y, and human capital each yield quasi-rents, while home-country market structure yields monopoly rents for the parent company. R&D for new product development may also be regarded

FIGURE 10.1 Relationships of Determinants of IFX and IFM to Three Received Theories on Foreign Direct Investment Abroad

| | Theories | | |
Determinants	Transfer Pricing	Structural Market Imperfections	Internalization of Transactions Costs
1. Research and Development Intensity:			
(a) Capability to Innovate	+ IFX	O	+ IFX
(b) Capability to Develop New Products	+ IFM	+ IFX O IFM*	+ IFM
2. Capital Intensity of Production	+ IFX + IFM	— IFX + IFM	O
3. Skilled Workforce (human capital)	+ IFX + IFM	O	+ IFX + IFM
4. International Orientation of Parent Companies	+ IFX + IFM	O	+ IFX + IFM
5. Consumer Products Marketing Skills or High Degree of Brand Loyalty	+ IFX + IFM	+ IFX — IFM	+ IFX O IFM
6. Minimum Efficient Scale of Production (relative to size of market)	+ IFX + IFM	— IFX + IFM	O
7. Market Seller Concentration	+ IFX + IFM	— IFX + IFM	O

Notes: * Applies to data sets with consumer goods only; positive otherwise.
+ = Positively related to IFT.
— = Negatively related to IFT.
O = No predicted relationship to IFT.

as a market-structure dimension. Each of these determinants is correlated positively between parent MNEs and their foreign affiliates (Connor 1977). As rents accrue from either home-country or host-country operations, the need for a threshold level of IFT arises. Moreover, the international experience of MNEs is hypothesized to provide administrative structures necessary to implement a successful IFT strategy.

Figure 10.1 also summarizes predicted relationships for an MNE exercising market power. Sugden (1983) presents his model specifically for IFM, but suggests that IFX determinants will be the opposite of the IFM case. His model assumes that there are N profit-maximizing firms producing a homogeneous product either in the home market, overseas, or both (the last are MNEs). Imports are from two sources: arm's length imports from independent overseas firms or intra-firm imports through domestically based MNEs. He also incorporates a fourth kind of firm, MNEs that do not engage in international trade of any kind. Several simplifying assumptions are made about supply curves, demand curves, and transfer costs associated with international trade. The model is oligopolistic because there are a finite number of firms. Each firm makes conjectures about the response of total domestic industry sales (Q) to a marginal change in output by any other firm (q_i) supplying that market. Ignoring exports and inventories, $Q = D + M + IFM$, where D is total domestic production, M is imports by the non-MNE "overseas" firms, and IFM is intra-firm imports by MNEs.

From these assumptions, Sugden (1983) derives first order conditions which we transform into the Lerner index ($\mathcal{L}$):

$$\mathcal{L} = \frac{\alpha}{\eta} + \frac{(1-\alpha)}{\eta} \cdot T \cdot \frac{D + IFM}{D + M}, \tag{1}$$

where η is the absolute value of the industry's domestic own-price elasticity of demand, α is the industry's weighted average conjectural elasticity ($\alpha=1$ for monopoly and 0 for Cournot pricing), and T is a measure of concentration. T is high if (1) the domestic Herfindahl index (H) is high and MNEs have a high share of D or if (2) H is high and IFM is small relative to D.

Solving for IFM as a proportion of other domestic supply and rearranging terms, we obtain:

$$\frac{IFM}{D + M} = \left[(\mathcal{L} - \frac{\alpha}{\eta})(\frac{\eta}{(1+\alpha)T}) \right] - \frac{D}{D + M}. \tag{2}$$

From this we find that the percentage IFM is directly related to economic profits and inversely related to concentration T (or, under reasonable conditions, H). Percentage IFM should be positively related to capital intensity because empirical measures of the Lerner Index ($\mathcal{L}$) include quasi-rents from the ownership of capital assets. We also see that percentage IFM is higher when arm's length imports M are large relative to domestic production D, *ceteris paribus*.

One extension of Sugden's (1983) model covers the heterogeneous-products case. Advertising intensity signals the presence of barriers to entry due to product differentiation, but experienced MNEs will find these barriers less formidable than would foreign or domestic entrants. Thus, MNEs can more easily evade these host-country barriers through IFX than can comparable consumer-products manufacturers outside the home market that need to enter via arms' length exports. Alternatively, when U.S. MNEs purchase finished goods abroad (high A/S) using their wholesale-trade foreign affiliates for resale in the home country, they find interbrand rivalry to be a formidable obstacle. When importing producer goods (low A/S), IFM is an advantage. And we view R&D used to develop new consumer products as essentially supportive of the marketing function represented by advertising and other selling expenses; for producer goods, this argument has less force.

A second extension of Sugden (1983) involves economies of scale relative to market size (MES), a conventional proxy for technical barriers to entry. Because MES sets a lower bound to concentration, they are directly correlated. If the correlation is not too high, the regression coefficients of MES and H ought to take the same signs.

Figure 10.1 also lists hypotheses derived from internalization theory. Both Siddarthan and Kumar (1990) (hereafter S&K) and Benvignati (1990) agree that R&D intensity promotes both kinds of IFT because high-tech products are more complex, novel, and unfamiliar to consumers and producers who are buyers. Information asymmetry and buyer uncertainty are likely to generate market failures. In the case of high-tech components or capital equipment, loss of technological secrets inhibits trade between unrelated parties. The arguments for skill-intensive goods are much the same as for R&D intensity. In addition, S&K argue that high skill intensity is indicative of the need for after-sales service for consumer products or for training in the use of intermediate products. Similarly, high selling costs relative to sales would encourage MNEs to ship consumer goods to their own host-country distributors so that the parents can get feedback from consumers abroad. However, neither believes that advertising affects IFM. Benvignati (1990) reasons that advertising reaches buyers directly, making it unnecessary for importers to correct an information breakdown.[5] (If advertising in home countries spills over to host countries, then Benvignati's reasoning directly contradicts S&K).

Experience in international production is equivalent to having a corporate structure that can collect and process information on international markets efficiently. This prior investment will lower transactions costs of IFT relative to market transactions.[6] With respect to K/L, MES, and CR4, internalization theory appears to be neutral. Both internalization studies include the capital/output (or capital/labor) ratio, but expect it to be unrelated to IFT. Like MES, capital intensity should foster *general* export intensity or affect the *location* decision, but neither appear to affect IFT. Neither study considers parent-industry concentration related to transactions costs.

Data Definitions

The variables used in our empirical tests are described in terms of U.S. parents and their affiliates abroad. Most data are developed from the benchmark or annual surveys of U.S. direct investment abroad (BEA 1982-1989). The benchmark surveys are universe data, and the annual surveys are large stratified samples. In the case of IFT undertaken by nonbank U.S. affiliates of foreign MNE, we used data from benchmark or annual surveys of foreign direct investment in the U.S. (1980-1988). Details are given in Connor and Wang (1993).

The *dependent variables* are the following:

IFX: intra-firm exports to all foreign affiliates as a proportion of total exports of U.S. parents (index no. 2, Table 10.1); and

IFM: intra-firm U.S. imports from foreign affiliates as a proportion of total imports of U.S. parents (index no. 2).

The *independent variables* are:

RD/S: research and development expenditures for U.S. parents (by the parents themselves or by other firms under contract) as a proportion of total sales of U.S. parents.

SKIL: in most cases we used the average wage per employee of U.S. parents (PSKIL) or affiliates (ASKIL). However, for the models involving food industries, we used the ratio of highly-skilled employees (scientists, engineers, and managers) to the total labor force in each U.S. industry.

K/S: the value of net property, plant, and equipment of U.S. parents as a proportion of total sales of U.S. parents. In some models we substituted (K/L): the ratio of total assets to total compensation paid by U.S. parents.

SELL: the ratio of selling, general, and administrative expenses of U.S. parents (PSELL) or affiliates (ASELL) to total sales.

INTL: total sales (or assets) of foreign affiliates as a proportion of total sales (or assets) of their U.S. parents.

OWN: total assets (FDI position) owned by the non-U.S. parents
        divided by the total assets of the U.S.-based affiliates of
        the foreign parents in a given industry.

A/S: weighted average of U.S. advertising expenditures in an
        industry as a proportion of total industry receipts.

MES: the ratio of average plant size to industry size for top
        50% of plant size distribution -- weighted 4-digit SIC
        industry data from the Census of Manufactures (1982,
        1987).

CR_4:   the 4-digit SIC industry four-firm concentration ratio,
        weighted by industry shipments value, from the Census
        of Manufactures (1982, 1987).

POL: a dummy variable taking the value of one for industries
        with high pollution abatement costs and zero for others
        as used by Siddharthan and Kumar (1990).

Results

We describe below regression models fitted against data for 1982 FDI manufacturing abroad. We test whether the results are robust using similar 1989 data. Then we examine the determinants of IFT in the case of *food* manufacturing.

FDI Abroad in 1982

Results are shown in Table 10.2 for exports (IFX) and Table 10.3 for imports (IFM). These models explain from 64 to 74 percent of the variance in IFX across the 32 industries, an improvement in goodness of fit of up to 14 percentage points compared to Siddarthan and Kumar's (1990) results. In this and succeeding tables, we first present models based on the internalization variables, and then we try to gauge the influence of additional market structure variables (A/S, MES, and CR4). Parental research and development intensity (PRD/S) is positive and highly significant in all models; the same can be said of the intensity of selling expenses (PSELL). Somewhat surprising was the fact that industry advertising intensity (A/S) was not a good substitute for parents' selling expenses. Doubtless, the highly diversified nature of the parent MNEs is partly responsible for this result.[7] Capital intensity (PK/L) was consistently negative in all models. Our corrected measure of international orientation (INTL) was a much larger and more significant determinant of IFX than previous researchers have found.

When we include PRD/S and two other market structure variables (MES, CR4), the parent company's employees' skill levels no longer affects IFX. (PSKIL's coefficient was close to zero). As is well known from U.S. studies, wages are correlated with imperfect competition in domestic markets (Scherer and Ross 1990). Thus, SKIL appears to be a good proxy for oligopoly market structure but a poor proxy for high transactions costs generated by high-cost human capital, an assumption made by previous researchers.   Economies of scale (MES) and concentration (CR4) both have negative effects on IFX, but they are highly correlated, so when both are included in the model neither is significant. The nonsignificance of A/S is also explained by its high correlation with PRD/S and PSELL.

In Table 10.3, regressions on 1982 intra-firm imports do not explain as much variation as do regressions for exports. (This is true for 1989 as well.) One reason for the poorer fit is our decision to omit the pollution-policy dummy variable (POL) from Siddharthan and Kumar (1990).[8]

IFM is strongly, positively associated with R&D intensity, just as are exports.   In our simplest model (3.1) R&D is the only significant determinant of IFM. A second significant factor is advertising intensity (A/S), our proxy for product differentiation.   Results indicate that average advertising intensity in the U.S. market strongly discourages intra-firm imports. Conversely, one might say that U.S. multinationals primarily import from their affiliates intermediate goods (parts, components, raw or semi-processed materials) or capital equipment. Intense advertising acts as a barrier to successful importation of foreign affiliates' finished consumer goods for resale in the U.S. market. Interestingly, although the two are highly correlated, A/S does a much better job of explaining IFM than does the parent's overall selling costs, a measure more of *enterprise* differentiation than of *product* differentiation. Note how different a result this is than for intra-firm exports, where U.S. multinationals use their enterprise image to foster such exports.

In one model (3.3) that omits selling costs, we show that scale economies inhibit the share of imports that is intra-firm, as was true for IFX. The products being imported from foreign affiliates are primarily intermediate or capital goods compatible with smaller-scale production processes. This is consistent with the finding that capital intensity is generally unrelated to IFM.   Work on international comparisons of market structures generally supports the finding of Bain (1966) that smaller national economies have smaller plant scales for the same industries. Since foreign affiliates operate in smaller national economies

TABLE 10.2 Regression Results Explaining Intra-Firm Exports, U.S. Foreign Direct Investment Abroad, 32 Manufacturing Industries, 1982

Equation	Intercept	PRD/S	PK/L	PSELL	INTL	PSKIL	A/S	MES	CR4	General Statistics R^2	$\overline{R}^2$
				Regression coefficients (t-test)							
2.1	0.023	2.44** (2.09)	-0.036** (-1.77)	0.79** (2.22)	0.60* (4.01)					0.70	0.65
2.2	0.158	3.12* (2.56)	-0.038*** (-1.54)		0.59* (3.53)		0.48 (0.34)			0.64	0.59
2.3	0.091	2.35*** (2.05)	-0.033*** (-1.67)	0.65** (1.79)	0.71* (4.23)				-0.002*** (-1.36)	0.72	0.66
2.4	0.052	2.48** (2.24)	-0.033*** (-1.67)	0.73** (2.14)	0.65* (4.47)			-0.021** (-1.99)		0.74	0.69
2.5	-0.110	2.19** (1.82)	-0.036** (-1.73)	0.97** (2.34)	0.55* (3.34)	0.004 (0.86)				0.71	0.65
2.6	-0.070	2.30** (1.94)	-0.032*** (-1.61)	0.91** (2.19)	0.57* (3.02)	0.004 (0.73)		-0.025 (-1.26)	0.0006 (0.24)	0.74	0.67

Note: Superscripts *, **, and *** represent significantly different from zero at the 1, 5, and 10 percent levels, respectively.

than do U.S. parents, imported inputs may incorporate scale features complementary with scales of production not found in the U.S.

Possessing a highly skilled workforce has no systematic effect on IFM. This is not surprising, as PSKIL and RD/S are conceptually quite close and positively correlated ($r = 0.30$). But the fact that international orientation and experience have no influence on IFM *is* a surprise. Apparently, the organizational experience acquired by parent firms from operating multinationally is useful when exporting to affiliates, but such expertise cannot be directed toward stimulating intra-firm imports. Other than the fact that the mix of goods exported is different from that imported intra-firm, it is difficult to imagine why this is so.

Finally, we hypothesized that IFM might operate in a fundamentally different way for consumer goods than for goods purchased by other businesses. Splitting a data set in this fashion is a common procedure in empirical industrial-organization studies (Scherer and Ross 1990). Merely including a consumer-good dummy variable was not successful. Results for the 15 consumer-goods industries are shown in equations 3.5 and 3.6 and for the 17 producer-goods industries in equation 3.7. Both subsets fit the data nearly as well as the all-manufacturing regressions. As expected, A/S is a better explanatory factor for consumer-goods than for producer-goods industries, whereas for PRD/S the reverse is true. Except for advertising intensity, market-structure factors do not affect IFM once we split the data.

Are the cross-sectional results for 1982 sensitive to time period? To answer this question, we ran regressions using the 1989 benchmark survey of FDI abroad. Results for IFX indicate some degree of time-wise sensitivity (Connor and Wang 1993). Perhaps the biggest changes are the much-diminished role for advertising intensity or selling costs and the much bigger role scale economies and concentration have in explaining intra-firm exports. Overall, market-structure variables yield stronger explanatory power than internalization variables. In 1989, MNEs in high-tech, concentrated industries with smallish plants were the most prone to export intra-firm. On the other hand, international orientation has a positive, but weaker effect on IFX in 1989. R&D intensity and capital intensity both stimulate 1989 IFM, whereas U.S. advertising intensity acts as a barrier to intra-firm importing. However, unlike IFX, other aspects of market structure only weakly affect IFM. The positive effect of K/Y on IFM in 1989 is another major difference.

TABLE 10.3 Regression Results Explaining Intra-Firm Imports, U.S. Foreign Direct Investment Abroad, 32 Manufacturing Industries, 1982

Equa-tion	Inter-cept	PRD/S	PK/L	PSELL	INTL	PSKIL	A/S	MES	CR4	Consumer Dummy	R²	R̄²
					Independent Variables						General Statistics	
					Regression coefficients (t-test)							
3.1	0.327	4.85* (2.78)	-0.013 (-0.44)	-0.69 (-1.13)	-0.003 (-0.01)	0.006 (0.92)					0.42	0.31
3.2	0.111	4.90* (3.19)	0.021 (0.68)		-0.05 (-0.21)	0.005 (0.86)	-4.60* (-2.56)	-0.033 (-1.27)	0.005 (1.29)		0.56	0.43
3.3	0.349	3.73** (2.11)	-0.028 (-0.094)		-0.07 (-0.26)			-0.054** (-1.95)	0.008** (2.05)	-0.13 (-0.07)	0.47	0.34
3.4	0.178	4.57* (2.64)	0.013 (0.38)		-0.03 (-0.10)	-0.003 (0.44)	-4.26** (-2.16)	-0.038 (-1.33)	0.006** (1.34)	-0.04 (-0.46)	0.56	0.41
					Consumer goods (N=15)							
3.5	0.125	8.02* (2.76)	0.12*** (1.70)				-11.29** (2.46)				0.49	0.35
3.6	-0.058	2.40 (0.83)				0.020** (2.52)	-4.59** (-2.18)				0.59	0.48
					Producer goods (N=17)							
3.7	-0.389	4.58** (3.02)	0.01 (0.33)				-2.98*** (-1.36)				0.47	0.35

Notes: Superscripts *, **, and *** represent significantly different from zero at the 1, 5, and 10 percent levels, respectively.

TABLE 10.4 Pooled Regression Results Explaining Intra-Firm Trade, U.S. Foreign Direct Investment Abroad, Three Food Manufacturing Industry Groups, 1982-1989

Equa-tion	Dependent Variable	Inter-cept	PRD/S	PK/S	PSELL	INTL	PSKIL	A/S	MES	CR4	General Statistics R^2	$\bar{R}^2$
						Regression coefficients (t statistic)						
4.1	IFX	-0.39	34.30 (1.45)	0.025 (1.05)		0.522 (0.64)	0.009 (1.12)				0.40	0.25
4.2	IFX	0.45	34.05* (2.71)			-0.614 (-0.98)		7.98* (3.41)	0.295** (1.59)	-0.028* (-3.63)	0.74	0.65
4.3	IFX	0.08	30.34** (2.31)	0.067* (3.15)		0.409 (0.62)			0.723* (4.43)	-0.037* (-4.00)	0.72	0.63
4.5	IFM	-0.23	60.66* (3.73)	-0.0003 (-0.02)		-1.218** (-2.17)	0.008 (1.33)				0.55	0.43
4.6	IFM	0.48	41.13* (4.54)			-1.462** (-2.66)		5.23* (3.04)		-0.015* (-2.63)	0.66	0.58
4.7	IFM	0.25	50.39* (5.54)	0.044* (2.96)		-0.982** (-2.14)			0.473* (4.18)	-0.029* (-4.47)	0.79	0.72
4.8	IFM	0.34	53.37* (5.16)			-1.543* (-2.98)		3.35** (1.75)	0.273** (1.79)	-0.021* (-3.32)	0.72	0.63

Notes: Superscripts *, **, and *** represent significantly different from zero at the 1, 5, and 10 percent levels, respectively.

Food Manufacturing FDI Abroad

There are only three food manufacturing industries for which FDI data are published across years. To overcome this problem we pooled eight years of available data (1982-1988) for a total of 21 observations.[9] The reader should recall that *food* IFT is only about half as high as *all manufacturing* IFT in 1987. Results are shown in Table 10.4.

The pooled regression results for the food manufacturing industries provide generally well-fitting and, unlike all manufacturing, consistent results for both exports and imports. For both types of IFT, R&D intensity and capital intensity stimulate intra-firm trade. (The latter result is sensitive to the inclusion of the highly correlated variables PSKIL, CR4, MES, and A/S.) As before, PSKIL is highly correlated with R&D intensity. The only significant departure from parallel import-export findings is the greater statistical significance of INTL in the IFM equations (4.5-4.9). The negative sign for INTL for IFM in the food industries is puzzling as it is contrary to all the previous cross-sectional results for all manufacturing, where INTL was non-negative.

Perhaps the most striking finding for the food industries is the consistently strong results for the three market-structure variables, A/S, MES, and CR4, even though they are highly collinear with each other. Not only are the three dimensions of market structure markedly better explanatory variables, but their signs are the *opposite* of the all-manufacturing results explained above. *Food MNEs located in less concentrated industries, producing highly differentiated, high-tech foods in large scale plants are the most likely firms to engage in intra- firm trade (of both types).* The structural sources of IFT in food products are notably different from those of other manufactured products.

Conclusions

Our principal purpose has been to assess the relative explanatory ability of three competing concepts of the drivers of intra-firm trade. To assess explanatory power, we developed two criteria. The first uses goodness of fit (R^2) to evaluate alternative explanatory models, where each model variant incorporates the independent variables peculiar to each theory. In the case of IFT models, we were able to identify unique variable combinations only for the internalization and market-structure theories. Because we analyzed five samples for both IFX and IFM, ten comparisons were possible.[10] In nine of the ten cases, the market-structure models outperformed internalization-based models, usually by a significant margin. On average, R^2 for a market-structure regression was 15 percent higher than for a comparable internalization model.

The second measure is based upon predictive *accuracy* of alternative regression models. Accuracy refers to the proportion of correctly predicted signs of regression coefficients rather than goodness of fit. Counting the proportion of correct signs is an admittedly crude method, but more formal statistical tests require that theories be nested. One clear pattern that emerges is that market-structure theory yields correct predictions on average 71 percent of the time — not perfect, but fairly satisfactory. Surprisingly, internalization theory produced predictions that were equal to or lower than what would result from a random outcome (i.e., 33 percent). The transfer-pricing explanation was only slightly better than chance would lead one to expect.

Two other findings stand out. First, one of the theories provided a good basis for predicting IFT when the sample consisted of the U.S. affiliates of non-U.S. MNEs. The relative newness of foreign direct investment in the United States (the majority made since 1980) may imply that we were observing a disequilibrium situation. U.S. FDI abroad is far more mature, permitting a higher degree of U.S. parent-foreign affiliate integration than is likely in the U.S. affiliate case. Second, the observed differences in determinants of IFT between U.S. *food* MNEs and other manufacturing MNEs is striking and puzzling. Sugden (1983) expected the structural determinants of IFM to be the opposite of those of IFX, but U.S. food manufacturers display identical signs for both flows. Domestic market sales concentration, for instance, depresses intra-firm food trade in both directions, whereas economies of scale stimulates such trade symmetrically. General manufacturing IFT responds differently to variation in concentration and scale economies, yet there is nothing in Sugden's (1983) model that would lead one to expect disparate results between food and the rest of manufacturing.

Notes

1. There is more extensive literature on intra-industry trade than on IFT. The two appear not to be correlated across industries (Helleiner 1979). However, IFT *is* closely related to "MNE-related" trade, in which one or more of the parties is a MNE. Kravis and Lipsey (1992) use the ratio of U.S. MNE-related trade to total world exports as an indicator of international competitiveness or "revealed comparative advantage".

2. The U.S. publishes IFT data for up to 62 industry groups.

3. In naming affiliates we follow the convention of the U.S. Bureau of Economic Analysis. U.S.-owned affiliates located outside the United States are

called *foreign* affiliates, whereas the affiliates of non-U.S. owners with their assets in the United States are termed *U.S.* affiliates.

4. These data do not permit one to distinguish between trade with the parent and trade with the parent's affiliates in third countries.

5. S&K's argument is illogical. IFM, they assert, is mostly intermediate products, so selling effort is likely to be absent. This is equivalent to saying that A/S or SELL has no variability.

6. S&K indicate in their equation (2) that INTL is expected to be unrelated to IFM, but their discussion does not confirm this. Benvignati is silent on the issue.

7. MNEs are large companies whose activities are typically spread across two or more of the 32 manufacturing industry groups represented. However, in the absence of more information, we assumed that the A/S of the MNE's *principal* industry was representative of all its selling effort.

8. Siddharthan and Kumar (1990) argue that MNEs transfer the production of dirty inputs to LDCs for reimportation to the home country. While the extra costs of pollution abatement may stimulate *foreign direct investment* from the United States to developing economies with ineffective environmental regulations, we fail to understand why intra-firm *trade* would be stimulated.

9. Prior to 1982 and after 1989, BEA no longer separated parent selling and administrative costs from other costs of sales. Thus, we could not calculate PSELL for these years.

10. For results using a 1987 cross-section of U.S. manufacturing affiliates and a 1980-1988 time-series of U.S. food manufacturing affiliates, see Connor and Wang (1993: Tables 6 and 7).

References

Bain, J. S. 1966. *International Differences in Industrial Structure*. New Haven, CT: Yale University Press.

Benvignati, A. M. 1990. "Industry Determinants and 'Differences' in U.S. Intrafirm and Arms'-length Exports." *Review of Economics and Statistics* 72:481-488.

Brander, J. and P. R. Krugman. 1983. "A Reciprocal Dumping Model of International Trade." *Journal of International Economics* 15:313-321.

Bucks, D. R. 1991. "Will the Emperor Discover He Has No Clothes Before the Empire is Sold?" *National Tax Journal* 44:311-314.

Bureau of the Census. 1982 and 1987. *Census of Manufactures: Industry Series* Washington, DC: U.S. Department of Commerce.

BEA (Bureau of Economic Analysis). 1993. *U.S. Direct Investment Abroad* 1989 and previous years. Washington, DC: U.S. Department of Commerce.

_____. 1991. *Foreign Direct Investment in the United States* 1988 and previous years. Washington, DC: U.S. Department of Commerce.

Casson, M. C. 1987. *The Firm and the Market*. Cambridge, MA: MIT Press.

Caves, R. E. 1971. "International Corporations: The Industrial Economics of Foreign Direct Investment." *Economica* 38:1-27.

Coase, R. 1937. "The Nature of the Firm." *Economica* 4:386-405.

Connor, J. M. 1977. *The Market Power of Multinationals.* New York, NY:Praeger Publishers.

_____. 1983. "Determinants of Foreign Direct Investment by Food and Tobacco Manufacturers." *American Journal of Agricultural Economics* 65:395-404.

Connor, J. M. and K. Wang. 1993. *The Determinants of Intra-Firm International Trade,* Staff Paper SP 93-11. West Lafayette, IN: Department of Agricultural Economics, Purdue University.

Dunning, J. H. 1979. "Explaining Changing Patterns of International Production: In Defense of the Eclectic Theory." *Oxford Bulletin of Economics and Statistics* 41:269-295.

Ethier, W. J. 1986. "The Multinational Firm." *Quarterly Journal of Economics* 101:805-833.

Graham, E. M. 1990. "Exchange of Threat Behavior Between Multinational Firms as an Infinitely Repeated Noncooperative Game." *International Trade Journal* 4:259-277.

Grubel, H. G. and P.J. Lloyd. 1975. *Intra-Industry Trade.* New York, NY:Wiley.

Helleiner, G. K. 1979. "Transnational Corporations and Trade Structure: The Role of Intra-Firm Trade," in H. Giersch, ed., *Intra-Industry Trade.* Tübingen:J.C.B. Muhr, P. Siebeck.

Helleiner, G. K. and R. Lavergne. 1979. "Intra-Firm Trade and Industrial Exports to the United States." *Oxford Bulletin of Economics and Statistics* 41:297-311.

Henneberry, S. and J. Russell. 1989. "Transfer Pricing in Multinational Firms: A Review of the Literature." *Agribusiness* 5:121-137.

Hymer, S. H. 1960. *The International Corporations of National Firms: A Study of Direct Foreign Investment.* Cambridge, MA:MIT Press.

Kindleberger, C. P. 1969. *American Business Abroad* New Haven, CT:Yale University Press.

Knickerbocker, F. 1973. *Oligopolistic Reaction and the Multinational Enterprise.* Boston, MA:Harvard Graduate School of Business Administration.

Lall, S. 1978. "The Pattern of Intra-Firm Exports by U.S. Multinationals." *Oxford Bulletin of Economics and Statistics* 40:209-222.

Lawrence, R. Z. 1990. "How Open is Japan?" in P.R. Krugman, ed., *Trade with Japan: Has the Door Opened Wider?* Chicago, IL:University of Chicago Press.

Rugman, A. M. 1980. "Internalization as a General Theory of Foreign Direct Investment: A Re-Appraisal of the Literature." *Weltwirtschaftliches Archiv* 116:365-379.

Scherer, F.M. and D. Ross. 1990. *Industrial Market Structure and Economic Performance.* Boston, MA:Houghton-Mifflin.

Siddharthan, N. S. and N. Kumar. 1990. "The Determinants of Inter-Industry Variations in the Proportion of Intra-Firm Trade: The Behavior of U.S. Multinationals." *Weltwirtschaftliches Archiv* 126:581-591.

Sugden, R. 1983. "The Degree of Monopoly, International Trade, and Transnational Corporations." *International Journal of Industrial Organization* 1:165-187.

United Nations. 1988. *Transnational Corporations in World Development: Trends and Prospects.* New York, NY:United Nations Centre on Transnational Corporations.

Vaitsos, C. V. 1974. *Intercountry Income Distribution and Transnational Enterprises.* London:Oxford University Press.

Vernon, R. 1966. "International Investment and International Trade in the Product Cycle." *Quarterly Journal of Economics* 91:190-207.

Williamson, O. E. 1975. *Market and Hierarchies: Analysis and Antitrust Implications.* New York, NY:Collier Macmillan Publishers.

_____. 1989. "Transactions Cost Economics," in R. Schmalensee and R. D. Willig, eds., *Handbook of Industrial Organization.* Amsterdam:North-Holland.

11

Foreign Investment Strategies of U.S. Multinational Food Firms

Michael R. Reed and Yulin Ning

Introduction

U.S. food companies are competing in an increasingly global market. In 1989 their food sales in foreign markets reached $87.6 billion and these sales are growing rapidly – from $62.8 billion in 1987 (BEA, U.S. Bureau of the Census 1990). U.S. food markets are saturated, but there is significant growth potential overseas because of faster income and population growth.

There are many ways that a firm can gain foreign markets. They can export, invest directly, license, or form a joint venture. This paper focuses on foreign direct investment (FDI) by U.S. food firms. Specifically, the paper investigates decisions on the degree of foreign direct investment by U.S. food firms and characterizes the basis for their business strategies. Strategies are analyzed in two ways: by case studies of five U.S. multinational enterprises (MNEs) and by a regression analysis with more firm observations (34).

Most large food companies produce overseas rather than export from the U.S. (Handy and Henderson 1994). Foreign affiliates of U.S. MNEs had food sales of $69 billion in 1989 (BEA 1993), much larger than their exports from the U.S. (only $16.6 billion for the same year). The question of why firms choose to become multinational has been studied for several decades in the economics literature, but little research has been done for the food industry. Some exceptions are Horst (1974) and Connor (1983, 1988). Other related works include Handy and MacDonald (1989), Handy and Henderson (1992), and Reed and Marchant (1992).

The decision on FDI is a firm-level decision (Caves 1971; Reed and Marchant 1992). For an individual firm to invest abroad, it must have some firm-specific advantages that enable it to overcome the intrinsic disadvantages caused by communication, linguistic, and cultural differences, and lack of knowledge about foreign markets. In most instances, licensing would be a much less cumbersome arrangement where the firm could capture most or all of the economic rents associated with its patents, copyrights or other intangible assets, without having to deal with a company which is quite distant from the home market. Nonetheless, given the same environmental factors, multinational firms favor a high control mode. Thus, multinational food firms are much more likely to invest directly than to export or license firm advantages.

Factors Leading to Globalization
by U.S. Food Manufacturers

It is useful first to discuss the reasons why U.S. food firms have become more interested in international markets in recent years. Many U.S. food processors face a U.S. market that is expanding at 1 percent or less per year in real terms. If these firms are to reward their stockholders with earnings growth, they must search out markets which are growing more rapidly. Many East Asian countries are growing at a 6-8 percent real rate per annum and there is a higher income elasticity of demand for most food items in these countries. Economic growth rates in Latin America and other parts of the world are also beginning to increase. This encourages firms to look to overseas markets for exports and investment.

The U.S. market is also becoming increasingly difficult for firms to have significant earnings growth, with new products constantly introduced and more brands available, food processor margins are not growing rapidly. Despite this competition, though, well-known brands are increasingly demanded in foreign markets. Many food brands which are well-known in the U.S. are highly demanded overseas also. U.S. food firms are discovering this and taking advantage of the economic rents associated with their brands.

Many countries are beginning to liberalize their domestic markets, lower trade barriers, and simplify investment rules. This makes these markets much more accessible to U.S. food processors. There is often a westernization of tastes which is occurring, which gives U.S. food processors a further advantage.

Finally, U.S. food firms are continuing to grow in terms of assets, sales, and other size indicators. These firms are seeking ways to diversify their markets, through widening product lines and geographic markets. Developing an international market for the firm's products will help reduce the risk due to general economic fluctuations, since economic cycles do not coincide among regions of the world.

Factors Leading to FDI by U.S. Food Manufacturers

The reasons for FDI are well-known and will be only mentioned briefly here. Based on Dunning's (1988) eclectic theory, a firm will engage in foreign production due to three conditions: ownership advantages of a firm, location advantages of a market, and internalization advantages of a firm and market. Unlike location advantages, which are external to the companies that use them, firm-specific advantages are internal factors. For an individual firm to invest abroad, it must have some firm-specific advantages that enable it to overcome the obvious disadvantages. These firm-specific advantages provide the first step in the explanation of FDI at the micro-level, which is the focus of this paper.

There are two basic types of FDI strategies: horizontal strategies, where a firm invests in a country to perform the same economic activity as in the home country; and vertical strategies, where a firm invests in a country to perform another activity in the marketing chain. Mixed global strategies are a combination of horizontal and vertical strategies where a firm uses FDI for global sourcing and market development (Helpman and Krugman 1985). Vertical FDI allows the investing firm to take advantage of lower costs in other countries. Wages, raw product prices, taxes and other costs of production may be lower in foreign countries than in the U.S.

Horizontal investment is the most common type of FDI for U.S. food firms (Reed and Marchant 1992). The most obvious reason for horizontal investment is to circumvent trade barriers and lower transportation costs. Exchange rate risk is also lowered through FDI and products are more easily tailored to specific demands within the country if production facilities are within the market.

No matter what the reason is for investment, FDI allows the U.S. food firm to capture its firm specific advantages through integrating into its foreign market. Patent rights, brand recognition, technological superiority, available capital, market power and other factors can be captured more fully through FDI. This paper focuses on measures of firm specific advantages that lead to entry into multinational status and to increased FDI.

In order to get a clear view of strategies of U.S. multinational food firms, the paper begins with five case studies to show how FDI strategies have varied by company. The analysis will focus on factors which have been important to the firm's global strategy. Then regression analysis is used to see how extendable those strategies are to other firms in the sample. The case studies offer a good opportunity to access the real business experiences in FDI. The regression analysis gives a more comprehensive appraisal of the intangible asset theory and firm-specific advantages.

FDI Strategies by Five Leading U.S. Food Manufacturers

This section outlines the foreign investment strategies of five major U.S. food processors. These companies were chosen because they were quite involved in FDI (each had more than 20 percent of their sales from foreign affiliates), had diverse product lines and quite different investment strategies. None of the firms export more than 2 percent of their U.S. output. In the analysis, particular attention is paid to the number and location of foreign processing plants, and sales, assets and earnings by region of the world.

The five processors are Borden, a diversified company with major investments in dairy, snack food and confectionery manufacturing and non-food investments in floor and wall coverings, adhesives and packaging materials; Campbell Soup, a food company with such well-known brands as Vlasic, Pepperidge Farms, Swanson, Godiva, Mrs. Pauls and V8; CPC International, which is a food manufacturer with many intermediate products, but with such brands as Mazola, Skippy and Knorr; Kellogg, which is a major breakfast food, dessert and snack food manufacturer with such brands as Poptarts, Leggo waffles and Mrs. Smiths pies; and finally Sara Lee, a diversified consumer products firm with frozen baked goods, processed meats and other food items, along with clothing (Isotoner, Hanes, L'eggs and Bali) and personal care products (Kiwi shoe polish). Table 11.1 shows 1985-90 growth in sales, earnings, assets and foreign processing plants by region for each company.

Sara Lee

Sara Lee is the largest of the five companies, with $12.3 billion in sales for 1991. Approximately 59 percent of those sales were food products, the lowest percentage among the five firms analyzed, and 29 percent of the food sales were by foreign affiliates. Sara Lee has been the most

TABLE 11.1 Sales, Earnings and Asset Growth Rates for the Five Firms, 1985-90 (Percent)

	Borden	Campbells	CPC	Kellogg	Sara Lee
U.S. Sales	-	35	32	47	25
European Sales	-	-	39	178	130
Total Sales	160	55	37	76	43
U.S. Earnings	-	21	286	46	53
European Earnings	-	-	72	177	308
Total Earnings	191	-24	105	78	97
U.S. Assets	-	17	138	120	116
European Assets	-	-	131	224	174
Total Assets	182	50	143	128	139%

Source: Annual Reports of Firms, 1985-90.

aggressive of the five companies, making major purchases or sales in almost every year (in 1988 alone there were 12 major purchases). Their number of foreign plants grew from 75 in 1978 to 248 in 1990. Most of their foreign affiliates are in Europe (179), but their largest growth in foreign affiliates has been in Asia and Australia.

Sales growth in the U.S. was lower for Sara Lee than any other company. Thus, the company has moved overseas and into non-food businesses to increase its growth. The firm, though, has seen its earnings grow faster than sales in all regions. The earnings growth was particularly rapid relative to sales in Europe, where earnings grew 308 percent in six years. The company has said that its non-food earnings have grown twice as fast as food earnings, yet its margins have increased in both areas. However, only in Europe has earnings growth outpaced asset growth. In general, Sara Lee is paying a high price for these companies relative to their sales and earnings thus far, but they are still experiencing substantial earnings growth.

Borden

Borden was the second largest of the five in 1991, with sales totalling $7.2 billion. Food products accounted for 74 percent of sales during that year and 22 percent of those food sales came from foreign affiliates. It had 47 foreign processing plants in 1976, compared with 111 in 1990. Most of that affiliate growth has been in Europe (42 plants were added) and Latin America (11 plants were added). There were two years when Borden made substantial acquisitions: 1979, when they purchased a number of food companies which increased their European presence by 17 plants and their Latin American presence by 4 plants; and 1989, when they acquired 17 European plants that included food and non-food processing plants.

Since 1985, earnings growth (191 percent) has exceeded sales growth (160 percent) and asset growth (182 percent). Unfortunately, we couldn't find these data by region of the world. Thus, Borden has done well since 1985, realizing the highest overall earnings growth of the five companies, but it is unknown whether this growth has come from U.S. or foreign operations, and whether it is from their food or non-food businesses. However, given their aggressive purchases of overseas plants in recent years, it is likely that foreign processing activity has played a large role in this growth.

Campbell Soup

Campbell Soup is the third largest company analyzed with sales totalling $6.2 billion in 1991. All of those sales were food products and 28 percent were from foreign affiliates. Campbells has been acquiring firms on a regular basis, but only acquisitions since 1985 have had an impact on their number of foreign processing facilities. Between 1985 and 1990, they concentrated on European expansion, with the number of processing facilities increasing from 12 to 33. However, this European expansion has been a disaster, with losses mounting each year since 1989. Management has said that they simply paid too much for those assets.

It appears that Campbells has tried to increase its overall growth by expanding overseas. Their assets have grown much faster than sales, especially in Europe, yet their foreign affiliate numbers have been rather stable since 1985. Slow U.S. sales, which only grew 35 percent over the 1985-90 period, at least generated earnings growth, which is more than the European assets did. One would have to judge Campbell diversification/growth strategy in Europe as a dismal failure thus far.

CPC International

CPC International had sales of $6.2 billion during 1991 and 94 percent of those sales were food products. CPC had the highest percentage of sales accounted for by foreign affiliates (60 percent), but it also had the smallest overall sales growth of the five firms. Note that it is less associated with branded consumer products than the other four companies (CPC concentrates quite a bit in wet corn milling). It appears that CPC is interested in diversifying it asset base and is trying to enter faster growing markets. It has acquired a number of firms (especially since 1987), but its number of foreign processing facilities has only grown from 83 in 1978 to 86 in 1990.

CPC has reduced its number of European processing facilities (from 42 in 1978 to 33 in 1990), but its European assets have grown by 131 percent since 1985. They have obviously consolidated their European holdings into fewer, more efficient plants. In 1989, CPC purchased a number of European firms, increasing their assets in Europe from $1.28 to $1.97 billion, yet their number of European processing facilities fell by three. CPC has expanded its number of plants in Canada (from 3 to 8 between 1978 and 1990) and Latin America (from 23 to 30 between 1978 and 1990). These are likely to be higher growth markets, given CPC's product mix.

The earnings growth rate indicates that CPC has entered into markets with a higher profit margin. Since 1985, earnings growth has exceeded sales growth in all markets. Earnings growth has been especially high in the U.S. (286 percent from 1985 to 1990). Asset growth has been around 140 percent in the U.S., Europe, and throughout the rest of the world.

Kellogg

Kellogg's worldwide sales were $5.8 billion in 1991, all in food products. Foreign affiliates account for 41 percent of the company's sales. Kellogg has mainly concentrated in breakfast food manufacturing over the years and has a significant market share in most areas of the world. In fact, its market share overseas is 45 percent versus 40 percent in the U.S.. In many parts of Latin America, Kellogg has a virtual monopoly (capturing 95 percent of the market in Mexico and 99 percent in Brazil). Its U.S. market is not growing very fast (45 percent between 1985 and 1990), so it has focused much of its attention on international markets.

Despite showing solid foreign sales gains (they have increased 660 percent since 1976), Kellogg has reduced its number of foreign manufacturing plants from 23 in 1976 to 18 in 1990. However, their asset increases (224 percent in Europe and 128 percent in total) indicates that they have been upgrading and expanding the capacity of the remaining facilities. The number of foreign processing plants peaked for Kellogg in 1982 at 26. It appears that most of the plants were simply shut down, rather than sold to another company. They haven't acquired many companies either, buying only one company between 1984 and 1990.

Kellogg is obviously looking toward Europe for the future, where it already dominates the UK market. However, it now faces more competition from the General Mills/Nestle joint venture. It is entrenched in Latin America (with 6 plants) and Canada (with 2 plants), but Europe is where sales, earnings and asset growth have been the strongest. One gets the feeling that they have slowly been taking advantage of the global recognition of the brand names and product lines. They don't buy other companies because there is less brand recognition associated with their competition and they have not substantially increased their product lines. They have purchased distribution system access and used their own plants, rather than buy plant and equipment through acquisitions.

Lessons from the Five Cases

From these five cases, there are three firms that have concentrated in food products with varying success. Campbells has tried to increase its international divisions, but has lost money in the process. They still must find a successful strategy or they will continue to have slow sales growth. Kelloggs has taken advantage of its brand identity and consolidated its assets into fewer plants with higher asset values. CPC is the final firm that is almost totally food-oriented, but it has moved into product lines that are more profitable, though the assets that it has purchased are rather costly. CPC is the most successful of the three food-oriented firms.

The two firms with a smaller percentage of sales in food seem to be doing better. Their earnings have increased faster than the totally food-oriented firms because they have diversified in product lines and geographic areas. Both firms are doing particularly well in Europe, but they seem poised for further expansion in Asia and Latin America.

Each of the five firms have continued to invest in food processing -- Campbells was the only firm where asset growth was less than 125 percent between 1985 and 1990. These firms have also increased their foreign sales more rapidly than their U.S. sales.

Regression Model Specification and Data

Past literature has focused on the role of technology, product differentiation, product diversity, firm size/scale, and firm concentration in FDI. Technological superiority is often a prerequisite for FDI (Bergsten, Horst, and Moran 1978; Lall 1980; Owen 1982; Grubaugh 1987). Firms with new and continually upgrading technologies and marketing skills many times want to assure their overseas markets by internalization of the advantages through FDI, rather than risk diffusion through licensing or exporting. Some food processing firms like Campbell Soup, Heinz, and others developed sufficient ownership advantages to expand their business both in the U.S. and abroad.

Yet most firms grow through mergers and acquisitions in order to fully exploit markets and capture profits, particularly where brand recognition and loyalty is not strong (such as grain milling). So concentration and product differentiation were ways to increase a firm's market power, and this can be accomplished through direct investment abroad. Product diversity is particularly related to the firm's risk management, and the higher the degree of a firm's product diversity, the more firm-specific advantages they have in other related production lines. Another variable, which should be included but has long been ignored, is considering that direct investment is only one strategy for MNEs to access international markets.

Regression analysis utilizing more firm data broadened the analysis to look at FDI behavior by U.S. MNEs. The first model explains whether a firm is an MNE (where an MNE has at least 10 percent of its assets in foreign countries), which is consistent with Grubaugh's (1987) idea that there is a large distinction between entry decisions and total FDI flows. The second model explains the degree of FDI by the firm. The independent variables in each model are identical and mostly stem from studies quite familiar to researchers in this area. These studies are identified with the variables included in the model.

Variables to measure technology (Bergsten, Horst, and Mason 1978; Lall 1980; Owen 1982; Grubaugh 1987), product differentiation (Caves 1971; Bergsten, Horst and Mason 1978; Lall 1980; Owen 1982; Pugel 1981; Grubaugh 1987) and firm size (Bergsten, Horst, and Mason 1978; Owen 1982; Grubaugh 1987) are included in the model. In addition, a variable to reflect whether the firm accesses international market through exports is included to assess whether exports and FDI are competitive marketing strategies.

The firm's technology is measured by three variables: research and development expenditures as a percentage of sales, the value of assets per employee, and sales per employee. Product differentiation is measured by the firm's marketing expenses as a percentage of total sales. Firm size is measured by total assets, while firm diversity is measured as the number of 4-digit SIC categories covered by the firm's output. All of these variables are expected to be positively related to FDI. The final variable is the firm's percentage of sales in foreign countries divided by the firm's percentage of assets in foreign countries (which will be called the export competitiveness index, for convenience). This variable is needed because the data set could not distinguish exports from affiliate sales for a country. If the coefficient on this final variable is negative, then exports and FDI are substitutes.

The general model specification follows:

$$FDI = f(RD, MKT, PRO, CA, DV, SIZE, EXP) \qquad (1)$$

where:

FDI is the measure of direct foreign investment (either entry or flow).
RD is research and development expenditures as a percentage of sales.
MKT is marketing expenditures as a percentage of sales.
PRO is sales per employee (a measure of productivity).
CA is assets per employee (a measure of capital intensity).
DV is the number of 4-digit SIC industries in which the firm is classified (a measure of diversity).
SIZE is total assets of the company.
EXP is the foreign sales as percentage of total sales, divided by foreign assets as a percentage of total assets (the export competitiveness index).
A firm subscript is implicit on each variable.

The data are based on a sample of 34 food processing firms listed in Handy and Henderson (1992). The data on these sample firms came from the Compustat and Global Worldscope databases.

The first model, which explains entry into MNE status, has a dichotomous dependent variable -- one if the firm is an MNE and zero otherwise. It is well-known that the linear probability model associated with this dependent variable is heteroskedastic with respect to the dependent variable, so a transformed generalized least squares estimation is performed.

The second model, which explains the degree or flow of FDI, has two dependent variables: foreign assets as a percentage of total assets and foreign sales as a percentage of total sales. The later variable minimizes problems in asset valuation. These two estimations are also performed using transformed generalized least squares to correct for heteroskedasticity associated with the dependent variable.

Results

The results of the multinational entry model are reported in column 2 of Table 11.2. Multinational food companies tend to be involved in many products (as witnessed by the significant coefficient on the diversity variable) which tend to be more capital intensive. This is consistent with the five cases analyzed earlier -- these companies must accumulate assets which will improve their competitiveness in various markets. They are not companies which have large research and development expenditures or involve highly productive processes. Further, large marketing efforts do not seem to be important in distinguishing MNEs from national-based firms. This is an interesting contrast to the five case firms. All of the five firms had well-established brand names which were supported by substantial marketing and promotion budgets.

TABLE 11.2 Results of the Models to Explain FDI Involvement

(1) Variable	(2) FDI Entry Model	(3) Foreign Assets/ Total Assets	(4) Foreign-Sales/ Total-Sales
Intercept	0.60** (2.32)	-3.78 (-0.29)	-40.49** (-3.29)
RD	-0.012 (-0.62)	-1.53 (-1.55)	-1.86* (-2.02)
MKT	0.009 (1.34)	0.39 (1.08)	0.73* (2.24)
PRO	-1.29 (-1.41)	34.53 (0.74)	-17.77 (-0.41)
CA	3.11** (2.57)	115.41* (1.87)	231.96** (4.03)
DIV	0.14** (4.13)	5.04** (2.92)	8.45** (5.25)
SIZE	-0.0044 (-0.63)	0.073 (0.21)	-0.067 (-0.20)
EXP	-0.57** (-4.38)	-21.36** (-3.24)	-13.32* (-2.17)
R-square	0.84	0.81	0.92

Notes: t-values are in parentheses.
* denotes significance at the 10 percent level.
** denotes significance at the 5 percent level.

The coefficient for the export variable indicates that exports are competitive with MNE status. Firms that have a high export competitiveness index are less likely to be an MNEs. In other words, firms with FDI in a country tend to service that market through affiliate sales rather than exports.

Columns 3 and 4 of Table 11.2 report the results of the regression to explain the degree of FDI. Capital intensity, product diversity and the export competitiveness index had the same signs in the flow equations as in the entry equation, and the coefficients on these three variables were also significantly different from zero in both flow equations. Again, this is similar to what was found in the five case studies. Access to capital and more product lines seemed very important to the five firms studied.

Marketing and research and development expenditures as a percentage of sales significantly affect the foreign sales percentage (column 4 of Table 11.2), but not the foreign asset percentage (column 3). Note that marketing expenditures were positively related to foreign sales percentage, while research and development expenditures were negatively related to foreign sales percentage. Marketing expenditures help the MNE use existing assets to maximize sales. However, high research and development firms must prefer to stay in their home-country base for fear of technology flowing to the host country.

Conclusions

The findings support Horst's (1974) contention that advertising and marketing by American firms are important in FDI success and that FDI is a diversification strategy, but advertising and marketing is not related to the FDI entry decision by firms. However, the results conflict with general (non-agricultural) FDI studies which find that productivity and firm size are major factors in decisions to invest (the entry equation) and in the level of investment (the flow equation). The sample of firms may account for some of this difference -- the sample may not adequately reflect the entire spectrum of U.S. food firms, especially in that it is oriented towards larger firms (which are much more likely to engage in FDI). Food firms may be different than other manufacturing firms (e.g., chemical and electrical equipment firms).

Technology does not seem to play a large role in FDI decisions for food processing firms. Firms with substantial involvement in research and development may prefer to remain in their home country and export the product, rather than risk the technology flowing to the host country through FDI.

Capital-intensive firms are more likely to engage in FDI, and the flow of FDI is usually larger than firms with lower capital requirements. This result, coupled with the lack of significance in the labor productivity variable, may indicate that FDI tends to relieve capital constraints for the host country, rather than provide a means to maximize returns to labor. Labor costs in most food processing enterprises may not be large enough to warrant FDI -- if labor productivity is not an important consideration in FDI, then wages may not be either.

These results have ramifications for investment flows by food manufacturers when trading blocs develop. For instance, they indicate that a North American Free Trade Agreement (NAFTA) may not have a large impact on vertical FDI flows to Mexico in food and kindred products. Wages may be lower in Mexico, but the labor component in food processing is low enough that wages differences do not matter. FDI in food processing may be more related to market access (i.e., circumventing trade barriers) and capital availability rather than technology, particularly since trade in intermediate food products is rather limited (Handy and Henderson 1992; Reed and Marchant 1992). This may mean that horizontal FDI in Mexico's food processing industry may increase rapidly if income growth continues.

The results consistently indicate that exports and FDI are substitute activities. This is consistent with the relatively small amount of trade in intermediate food products and the view that most FDI by U.S. food firms is horizontal in nature – it is a market access strategy. These findings are consistent with those less analytical results of Reed and Marchant (1992) and Handy and Henderson (1992).

Notes

1. According to Sheldon and Henderson (1990), at least half of those firms with international operations in the 120 world's largest food firms are engaged in product licensing. Certainly the degrees of involvement in export and direct investment are higher than that in licensing.

References

Bergsten, F., T. Horst, and T. Moran. 1978. *American Multinationals and American Interests*. Washington, D.C.: Brookings Institution.
Bureau of Economic Analysis (BEA). 1993. *U.S. Direct Investment Abroad* 1989 and previous years. U.S. Department of Commerce, Washington, D.C.

Caves, R.E. 1971. "International Corporations: The Industrial Economics of Foreign Investment." *Economica* 38:1-27.

Connor, J.M. 1983. "Determinants of Foreign Direct Investment by Food and Tobacco Manufacturers." *American Journal of Agricultural Economics* 65:395-404.

_____. 1988. *Food Processing: An Industrial Powerhouse in Transition*. Lexington, MA: Lexington Books.

_____. 1981. "Foreign Food Firms: Their Participation in and Competitive Impact on the U.S. Food and Tobacco Manufacturing Sector," in Johnson, G. and A. Maunder, eds., *Rural Change: The Challenge for Agricultural Economists*. Oxford: Oxford University Press.

Dunning, J.H. 1988. "The Electic Paradigm of International Production: A Restatement and Some Possible Extensions," *Journal of International Business Studies* 19: 1-31.

Grubaugh, S. 1987. "Determinants of Direct Foreign Investment." *Review of Economics and Statistics* 69:149-52.

Handy, C.R. and D.R. Henderson. 1994. "Foreign Direct Investment in Food Manufacturing Industries," in M. Bredahl, P. Abbott, and M. Reed, eds., *Competitiveness in International Food Markets*. Boulder, CO: Westview Press

Handy, C.R. and J. M. MacDonald. 1989. "Multinational Structures and Strategies of U.S. Food Firms." *American Journal of Agricultural Economics* 71:1246-54.

Helpman, E. and P.R. Krugman. 1985. "Market Structure and Foreign Trade." Cambridge, MA:MIT Press.

Horst, T. 1974. *At Home Abroad- A Study of the Domestic and Foreign Operations of the American Food-Processing Industry*. Cambridge, MA: Ballinger Publishing.

Lall, S. 1980. "Monopolistic Advantages and Foreign Investment by U.S. Manufacturing Industry." *Oxford Economic Papers* 32:325-48.

Owen, R. 1982. "Inter-Industry Determinants of Foreign Direct Investment: A Canadian Perspective," in A. Rugman, ed., *New Theories of Multinational Enterprise*. London: Croom Helm.

Pugel, T. 1981. "The Determinants of Foreign Direct Investment: An Analysis of U.S. Manufacturing Industries." *Managerial and Decision Economics* 2:220-28.

Reed, M. and M. Marchant. 1992. "The Global Competitiveness of the U.S. Food Processing Sector." *Northeastern Journal of Agricultural and Resource Economics* 22:61-70.

Sheldon, I.M. and Henderson, D. R.. 1990. "International Licensing of Branded Food Products." NC-194 Occasional Paper OP-15. Columbus, OH: Ohio State University.

U.S. Bureau of the Census. 1990. *Highlights of U.S. Export and Import Trade*. Series FT 990.

12

Industrial Determinants of International Trade and Foreign Investment by Food and Beverage Manufacturing Firms

Dennis R. Henderson, Peter R. Vörös,
and Joseph G. Hirschberg

Introduction

Regardless of a firm's nationality, foreign direct investment (FDI) and exports are prominent among the international strategies used by leading food and beverage manufacturing firms. Measured in terms of the value of international sales, for the world's leading firms FDI is roughly four times more important than exports (Henderson and Handy 1993). Yet, both foreign investment and export propensities vary widely among firms. For example, sales from foreign affiliates as a share of total corporate food and beverage sales range from zero to more than 60 percent [The Coca-Cola Company] for U.S. firms, and to more than 95 percent [Nestlé SA] for non-U.S. firms. Exports as a share of total sales range from zero to more than 30 percent [Riceland Foods] for U.S. firms and 55 percent [LVMH Moet-Hennessy Louis Vuitton] for non-U.S. firms.

Both theoretical and empirical literature cite characteristics of industrial organization as possible determinants of the extent to which firms pursue foreign investment and export strategies. For example, one theory of foreign direct investment holds that firms develop operations abroad in order to exploit intangible assets (Grubaugh 1987). Such assets include research and development (R&D), differentiated products, established loyalties with suppliers and customers, and unique brands and trademarks (Connor 1983). Product differentiation, along with other structural characteristics, such as the extent of vertical tie-ins, has also been shown to affect export propensities in U.S. food manufacturing industries (Henderson and Frank 1990).

The role of market power is somewhat more ambiguous. For example, Sheldon (1992) has argued that home market power could be export enhancing through dumping. By contrast, Porter (1990) has argued that home market rivalry associated with the absence of demonstrable market power is an important source of competitive advantage in foreign markets. Henderson and Frank (1990) found evidence supporting the latter view in the U.S. food processing industries. Pagoulatos and Sorensen (1975) found evidence supporting the former view among 88 U.S. manufacturing industries.

The purpose of this paper is to present results of empirical analysis of food and beverage manufacturing firms that test hypotheses relating a firm's dominance in its home market, product characteristics, and investment in intangible assets to export propensity (exports as a share of total sales), and FDI intensity (shipments from foreign affiliates as a share of total sales). Also examined is the extent to which exports and FDI are strategic behaviors -- i.e. they enhance profits. The analysis uses regressions which are run on an international sample of food and beverage manufacturing firms drawn from 2,512 annual observations reported in the *Global Company Handbook* (CIFAR 1992) from 1987 through 1990.

The paper is organized as follows: testable hypotheses are generated from an examination of conceptual and empirical literature, the analytical methods are described, empirical findings are presented, and conclusions and implications are drawn.

Expectations and Hypotheses

A robust literature on the interface between industrial organization and international trade theories has emerged since the early 1970s (see Jacquemin 1981; Caves 1985; and Pagoulatos 1992). The line of causation in much of this runs from trade to industrial organization. For example, imports impose a competitive discipline on domestic market performance.

Reversing the issue -- asking the question, what is the impact of industry structure and market behavior on international market performance -- results in less clear expectations. As Sheldon (1992) has noted, expectations depend on the benchmark one adopts for judging domestic and foreign market performance. The literature in international marketing, and to some extent, in industrial organization, recognizes pro-commerce type objectives (see Glejser, Jacquemin, and Petit 1980; Lyons 1989; Porter 1990; and Jasinowski 1991). In contrast, the literature in international economics addresses performance in terms of welfare maximization (see Bhagwati 1981).

In this study, following the literature that relates domestic market organization to international market behavior and performance, pro-commerce objectives are emphasized. By doing so, we do not imply a value judgment (e.g., the desirability of mercantilism), but take exports and sales from foreign affiliates to be relative indicators of the external market performance of an entity (firm, industry, country) compared to that of other entities.

Variants of industrial organization theory (IOT) offer some insight into the determinants of patterns of international trade. An early variant of IOT was based on the Structure-Conduct-Performance (SCP) paradigm. A basic tenet of SCP is that the way an industry is structured, the way the participants in it behave, and its performance are functionally related. The line of causation runs from structure to conduct and performance.

More recently this line of causation has come into question with the evolution of the New Empirical Industrial Organization (NEIO) paradigm (see Fort and Hallagan 1987; Marvel 1992; and Pagoulatos 1992). The NEIO paradigm holds that structure and performance are interrelated as a whole -- i.e., neither causes the other. It is also argued that structure and performance are jointly determined by underlying cost and demand parameters (see Shaked and Sutton 1987, 1990; Sutton 1991). As a result, more emphasis has been put on the role of market behavior in influencing both structure and performance.

Much of the industrial organization literature, both in the SCP and the NEIO veins, says little explicitly about international commerce. Yet some insights have evolved from both theoretical and empirical work.

Exports

In early work, White (1974) advanced a theoretical explanation that posits causality between domestic market structure and international trade. White's (1974) framework is a combination of monopoly and competition theory within the context of neoclassical trade theory and the small-country assumption that the domestic producer cannot affect world prices -- it is a price taker and maximizes profits accordingly. White's (1974) conclusions are as follows: (1) with trade barriers and the ability to price discriminate, dumping results in the monopoly case. Exports are greater than with a competitive domestic market structure. (2) With trade barriers and the inability to dump, a monopoly will export less than a competitive industry. (3) "If the domestic and foreign products are imperfect substitutes so the monopolist enjoys some market power even in export markets, then anything might happen, and one has to know the specific details of price elasticities and market positions before making any predictions" (White 1976, p. 1020).

Lyons (1981) argued that the impact of market power on exports is especially difficult to identify, except in the simplest forms of market structure. However, he was able to demonstrate theoretically that, in the presence of multinational firms, market power can lead to geographic market segmentation which could reduce world trade.

In an early attempt to test White's (1976) theoretical results Pagoulatos and Sorensen (1975) estimated regressions on 88 U.S. manufacturing industries. They found exports as a share of domestic shipments (export propensity) to be positively related to domestic market power, as well as to scale economies, product differentiation, and expenditures on research and development.

Rapp (1976) informally explored the relationships among firm size and export market development over 100 years in Japan. Rapp's (1976) conclusions included: (1) larger firms are more likely to export than smaller firms, (2) among firms that export, smaller firms are likely to be more export specialized, and (3) concentration increased with the export role of large firms over time.

Marvel (1980) simultaneously estimated the determinants of the composition of U.S. trade flows and price-cost margins. The major finding of the study was that managerial inputs, R&D, and demand considerations account for much of the difference between U.S. import and export bundles. He also found weak evidence of a negative effect of concentration on exports.

Glejser *et al.* (1980) examined a sample of 1446 Belgian firms to determine exports are related to domestic and foreign market structure. Their findings showed both concentration and product differentiation to have negative impacts on exports.

Koo and Martin (1984) focused on the interaction between strategic choice variables: different types of product differentiation, research and development, with scale economies, and market concentration in the determination of trade flows. The most significant relationships were found for market power (negative) and product differentiation. The two types of product differentiation showed different effects on exports: that stemming from complex nature of the product showed a positive effect on exports, but differentiation based on advertising had a negative effect.

Lyons (1989) estimated the determinants of import penetration and export propensity using 1968 and 1980 cross-sectional data from 111 3-digit industries in the UK manufacturing sector. The findings showed that advertising has a negative impact on export propensity, whereas economies of scale and research and development have positive effects. Conclusions about the effect of concentration were not clear cut.

Henderson and Frank (1990) studied the effect of a number of structural and behavioral characteristics on export propensity using regression analysis on 1982 cross-sectional data from 42 U.S. food manufacturing industries. The major findings were that exports were negatively related to home market concentration, advertising, cost of labor, trade barriers, vertical ties, and sales from foreign affiliates and positively related to scale economies and expenditures on research and development (R&D).

Lipsey (1991) tested the impact of labor, physical capital, human capital, advertising, and research and development on the international market exposure of U.S. manufacturing industries. He found human capital and R&D intensities to have positive effects, and labor intensity and advertising to have significant negative influences.

Sales from Foreign Affiliates

Turning to FDI, the theoretical literature on the determinants of FDI is somewhat more developed. There is a large literature addressing multinational enterprises, and another addressing the welfare consequences of FDI. Horstmann and Markusen (1989) have provided a model that bridges these two strands of thought. A key concept in the model is that of firm-specific assets, or the idea of ownership advantage. Ownership advantages can be factors such as "technical expertise gained through R&D..., managerial expertise..., and product reputation and identification through advertising" (Horstman and Markusen 1989, p. 46). Related concepts of firm-specific international advantage include scale economies in both production and distribution (Yu 1990) and experience (Kumar 1991). Firm-specific assets, or what may be called intangibles, have come to dominate theoretical explanations of foreign investment (see Ethier 1986; Casson 1987; Grubaugh 1987).

Helpman and Krugman (1985) have advanced an explanation of FDI that rests on two conditions: (1) the presence of product differentiation and economies of scale, and (2) some inputs (such as management, marketing, and product specific research and development) that are highly specialized and can be located in one country while serving product lines in another country. The latter are referred to as owner-specific advantages or firm-specific assets. Firms are assumed to make cost-minimizing location decisions for production facilities and "export" firm-specific factors. As prices for other than firm-specific factors differ across countries, multinational firms emerge, so FDI results.

Firm-related factors as determinants of FDI that have been examined in empirical research include the following: Yu (1990) reported a positive relationship between firm size and FDI, R&D, and FDI, and advertising intensity and FDI. Dunning (1981) reported a positive relationship between intangible assets and FDI. Specific to the food sector, Handy and Henderson (1992) showed a positive association between firm-specific assets, specialization in food products, and sales by foreign affiliates.

Whereas the literature on the relationship of trade to home monopoly, market power, and concentration is somewhat ambiguous, other than an implied positive relationship between firm size and foreign investment, little can be found in the literature that relates these dimensions of home market dominance to FDI. Several other factors that appear to be important in explaining trade patterns -- expenditures on R&D, advertising and product differentiation, and the quality of human capital, in the work on FDI -- appear to be largely subsumed under the broader classification of firm-specific or intangible assets.

Testable Hypotheses

A broad class of firm characteristics generally categorized as firm-specific or intangible assets, including such things as research and development, product differentiation, and technical expertise, appears to be a direct determinant of FDI. In addition, some components of intangibles (R&D and human expertise) appear to be positively associated with exports.

Other intangibles, such as product differentiation, have less clear associations with international commerce. For example, a few studies (Pagoulatos and Sorensen 1975; Koo and Martin 1984) have shown product differentiation associated with product complexity to be trade-enhancing, while several studies (Glejser et al. 1980; Koo and Martin 1984; Lyons 1989; Henderson and Frank 1990; and Lipsey 1991) have shown product differentiation associated with advertising (brand promotion) to reduce trade. After reviewing considerable literature, Hladik (1985) concluded that no specific relationship could be postulated between intangibles and export behavior, a conclusion subsequently borne out by her analysis. Herein, Hladik's (1985) convention is adopted.

Across a wide array of industries, scale economies appear to be positively related to trade (Pagoulatos and Sorensen 1975; Lyons 1989; Henderson and Frank 1990) and perhaps to FDI as a component of firm-specific advantages. Studies that have been specific to the food and beverage manufacturing industries (Henderson and Frank 1990; Handy and Henderson 1992) suggest that the degree to which firms specialize in the food sector enhances at least the trade component of foreign commerce.

The most uncertain expectations are those that posit relationships between a firm's dominance in its home market and its international market exposure. Yet when home market dominance is categorized, some fairly consistent themes emerge. A firm that holds a home monopoly would appear to have incentives to export, providing some form of price discrimination or dumping can be carried out (White 1974). Short of monopoly, exporting has been shown to be both positively (Pagoulatos and Sorensen 1975) and negatively (Lyons 1989) related to market power. When home market dominance is viewed in terms of market share or concentration, findings are more consistent; a firm's export propensity appears to be negatively related (Glejser *et al.* 1980; Marvel 1980; Koo and Martin 1984; Henderson and Frank 1990). No priors appear obvious when relating either market power or market share to FDI. Regarding firm size, one study suggested this is positively related to exports (Rapp 1976). However, this can be viewed as somewhat inconsistent with the findings regarding market share, presuming the two are positively correlated. Finally, firm size would seem to be positively related to FDI by definition.

Limitations in the data set that is available, described in the next section, impose constraints on the extent to which hypotheses can be empirically tested. Specifically, available data allow home market dominance to be expressed in just three ways: (1) a firm's aggregate share of its home country manufactured food and beverage market (AMS20), (2) firm size measured in terms of the value of sales of manufactured food and beverages (SIZE), and (3) net income as a percent of net total sales (NINT), a rough proxy for the relative size of a firm's price-cost margin as an indicator of market power. Data are also available to measure a firm's degree of specialization in food and beverage products (SPEC) measured as the value of manufactured food and beverage sales as a percent of total sales, and the breadth of a firm's product offering within food and beverages (PD20CT) as an indicator of product diversity or differentiation, specified as a count of the number of 3-digit SICs within which a firm produces a food or beverage product.

Based on previous conceptual and empirical work, and on the availability of data, hypothesized relationships are as shown in Table 12.1. Finally, the effect of international market exposure on firm profitability is investigated by examining the relationship between FDI, exports, and net income. Positive relationships are expected, or the international market behavior would not be observed.

Analytical Procedures

The multiple regression models used to explain FDI and exports are linear functions of the independent variables shown in Table 12.1 plus a set of dichotomous variables to account for idiosyncracies associated with year, industry, and country. Regression models were also used to estimate the effect of foreign sales (as exports, shipments from foreign affiliates, or both) on firm profitability, specified as returns on total assets (NITA).

As ordinary least squares (OLS) models, they are specified by the general form:

$$y_i = \beta' x_i + e_i \tag{1}$$

where x_i are the independent variables and e_i is normally distributed with mean of zero and variance of σ_e^2.

Firm-level data are used. Hirschberg, Dayton, and Vörös (1992) have described several sources of data on the structure, behavior, and performance of firms in the food and beverage manufacturing industries world-wide. One of the most consistent of these, in terms of the combination of number of firms and extent of information per firm, is the Global Company Handbook, published in machine-readable form by the Center for International Financial Analysis and Research (CIFAR 1992). A fairly complete set of annual investment and operating data could be extracted from CIFAR on 628 firms with food and/or beverage manufacturing operations, representing 41 countries, for the years 1987 through 1990. This generated potentially 2,512 annual observations.

CIFAR obtains its information primarily from annual reports issued by the firms in its population. Many firms do not report international operations as separable items. Thus, sufficiently detailed data are not present in all 2,512 observations to distinguish specific international

TABLE 12.1 Expected Determinants of Export Propensity and Foreign Direct Investment

Determinant	Export Propensity	Foreign Direct Investment
Home Market Dominance		
Aggregate Market Share (AMS20)	-	0
Firm Size (SIZE)	0	+
Price-Cost Margin (NINT)	-	0
Sector Specialization (SPEC)	+	0
Product Diversity (PD20CT)	+	+

market operations (i.e. export shipments from home country or shipments from foreign affiliates). Sample sizes of 105 and 312 were drawn for use in multiple regression models fit to export propensities (XP) and shipments from foreign affiliates (FDI), respectively.

Because of the relatively small number of observations that can be used for analysis of the phenomena of interest, the question of sample representativeness, or potential selectivity bias, demands attention. To be able to draw inferences from the available data it is necessary to ensure that the firms used in the analysis form a representative sample. In order to account for potential problems posed by using a non-representative sample, Heckman's (1979) two-step procedure for eliminating sample selectivity bias was employed. In that procedure, selectivity bias is defined as a specification error in the regression of interest. This error is then corrected by adding a new regressor which is computed from the parameters estimated in a first stage probit model.

The coverage of the CIFAR data for the variables used in the models described above is less than universal. For certain countries (e.g. Argentina, Austria, Belgium, and Japan) none of the firm-level data can be used to model either FDI or export behavior. At the same time 31 percent of the Indian firms provide export data and 26 percent provide sufficient data to model FDI. There are a number of explanations for these differences among firms and countries:

1. Reporting procedures for the annual reports (the equivalent of the 10K report required by the U.S. Securities and Exchange Commission) differ by country and may or may not include requirements to report the appropriate data.

2. For competitive reasons a firm may conceal details of its foreign operations.

3. A firm may not engage in FDI or export activities. Thus, the absence of data is not due to reporting error or purposeful concealment. However, there is no information to confirm the lack of international market activity.

4. The detail available in the CIFAR data base may be subject to cultural bias on the part of the data entry personnel. In the case of CIFAR most data entry is done in India.

In order to draw inferences from the available data, models were constructed to explain the presence or absence of pertinent data. For these first stage models, two probit models (Goldberger 1964) are estimated: one for the presence of export data and the second for the presence of FDI data. The coefficients of the probit model identify those factors that are most important in determining whether the observation is present in the sample of interest. In the second stage, following Heckman's (1979) method to eliminate the influence of sample selection bias, the estimated parameters from the probit model are used to form new independent variables which are then imposed in the XP and FDI regressions which, in turn, utilize observations that contain the international market phenomenon of interest. A complete description of this technique can be found in Maddala (1983, chapter 9).

The probit model for the tendency to report FDI is estimated with a linear model of the form:

$$d_i^* = \gamma' z_i + u_i \tag{2}$$

Where z_i is a set of characteristics for observation i and u_i is the error term with variance σ_u. In this case the regressors are total sales, net income, a dummy variable for country of origin, and a dummy variable for year. The values of the coefficients γ_i can be interpreted as the influences of these factors on the tendency to report FDI. However, only the value of d_i is observed, which is defined by:

$d_i = 1$ if $d_i^* > 0$ (FDI data are available),
$d_i = 0$ otherwise (FDI data are unavailable).

This model can be estimated using a number of computer programs. For this analysis, the LIMDEP computer package (Greene 1991) was used.

The second step in the Heckman (1979) procedure to account for sample selectivity bias is based on the result that the expected value of a random variable (r_1), that is distributed according to a truncated normal, is given as:

$$E(r_1 \mid r_2 > c) = \sigma_{12} \frac{\phi(c)}{1 - \Phi(c)} \tag{3}$$

where r_1 and r_2 are bivariate normally distributed random variables with σ_{12} as the covariance between the two variables and means equal to zero. The ratio $\phi(c)/(1-\Phi(c))$ is often referred to as the inverse Mills ratio, or the "failure rate" of the distribution.

Thus, the expected value of d is defined as a function of the inverse Mills ratio computed from the probit model under the assumption that e_i and u_i are bivariately normally distributed. This formulation is given as:

$$E(y_i \mid d_i = 1, x_i) = \beta' x_i + \sigma_{ue} \lambda(\gamma' z_i) \qquad (4)$$

where λ is the appropriate value for the inverse Mills ratio based on the probit analysis discussed above. Consequently, in the sample selection-adjusted models, the regression of y on x and λ will furnish an unbiased estimate of the parameter set:

$$y_i = \beta' x_i + \sigma_{ue} \lambda(\gamma' z_i) + v_i \qquad (5)$$

The representativeness of the sample is assessed by testing the parameter on the inverse Mills ratio against the hypothesis that no relationship exists between the tendency to report sufficient data for the analysis and the level of either FDI or XP.

In the analysis that follows, this procedure is applied to the models for XP, FDI, and profitability, and the results of the sample selectivity-adjusted models (SSMs) are presented in when we cannot reject the hypothesis that the parameter on $\lambda(\gamma' z_i) = 0$.

Results

Two variations of regression models, OLS and SSM, were estimated. Three versions of three models regressed potential determinants of international market exposure on XP and FDI, and single models regressed XP, FDI, and the combination of XP and FDI on profitability.

Our results reveal the extent to which international market activity is influenced by the hypothesized determinants: home market dominance as represented by AMS20, SIZE or NINT; intangibles; specialization; product diversity or complexity; and the extent to which international market behavior influences firm profitability. Furthermore, the significance of the parameter value for λ reveals the extent to which the findings are sensitive to the reporting habits of firms in the CIFAR data base.

Determinants of Export Propensity

Findings regarding the determinants of export propensity by firms with food and/or beverage manufacturing operations are shown in Table 12.2.

Three versions of the XP models are presented, each with a different specification of home market dominance: version A specifies home

TABLE 12.2 Parameter Estimates for the Model of Export Propensities

Variable	Version		
	A OLS	B OLS	C OLS
AMS20	-1.4415*		
SIZE		-0.0116*	
NINT			0.1751
SPEC	0.2017*	0.1931*	0.2076*
PD20CT	0.0186	-0.0306*	-0.0354*
Adj. R-square	0.538	0.547	0.501
F-value	7.049*	7.292*	6.218*
Sample Size	105	105	105

Note: * significant at the one percent level.

market dominance in terms of aggregate market share (AMS20), version B uses size of firm (SIZE), and Version C incorporates the price-cost margin proxy (NINT). Because the sample selection model did not show any evidence of sample bias, we only report the OLS results.

The results for all models show that a significant portion of the variation in the dependent variable XP is captured by variability in the specified independent variables, with F-values that rejects the hypothesis that the non-intercept terms are equal to zero and relatively high adjusted R-square values for an analysis of cross-section data.

The results are generally consistent with expectations. An inverse relationship between export propensity and a firm's dominance in its home market appears to be confirmed, with significant and negative coefficients for both AMS20 and SIZE and an insignificant coefficient for NINT. Given priors for a negative relationship for AMS20 and NINT and an ambiguous relationship for SIZE, this suggests that, considering the nature of the available data, AMS20 is the superior of the three measures of home market dominance. As hypothesized, the extent to which firms specialize in food and/or beverage manufacture (SPEC) is positively related to exports. The interpretation of the parameter estimates for PD20CT, which tend toward the significantly negative, is consistent with expectations based on the interpretation of this variable as a proxy for product diversity or differentiation.

Determinants of Shipments from Foreign Operations.

Findings regarding the determinants of the share of a firm's total value of shipments originating from a foreign food and/or beverage manufacturing affiliate (FDI) are shown in Table 12.3.

As with the XP analysis, three versions of the FDI models are presented, each with a different specification of home market dominance: Version A using AMS20, Version B using SIZE, and Version C incorporating NINT.

In this case only the SSM variants are presented, because the parameter estimates for λ are statistically significant in the FDI models. Thus, the relationships fitted to explain FDI are affected by the probability that a firm reports FDI data. As Heckman (1989) notes, the value of the inverse Mills ratio $\lambda(\gamma'z_i)$ is a monotonically decreasing function of the probability that the observation will be in the sample (Prob($d^*>0$)). Thus, the actual occurrence of FDI is negatively related to the tendency to report FDI data. That is, we find that:

$$\frac{\partial FDI}{\partial d^*} < 0$$

indicating that as a firm has a higher tendency to report FDI, its level of FDI is less than otherwise.

TABLE 12.3 Parameter Estimates for the Model of Foreign Operations

Variable	Version		
	A SSM	B SSM	C SSM
$\lambda(\gamma'z_i)$	-0.0786*	0.1345*	0.1107*
AMS20	-1.7990*		
SIZE		0.0156*	
NINT			0.2020**
SPEC	0.0364	0.0632**	0.0116
PD20CT	0.0555*	0.5212*	0.0683*
Adj. R-square	0.340	0.378	0.331
F-value	6.183*	7.264*	5.910*
Sample Size	312	312	312

Note: *, ** significant at the one and ten percent levels, respectively.

The results for all models show that a significant portion of the variation in FDI is captured by the variability in the specified independent variables, with significant F-values. Adjusted R-square values, however, are somewhat lower than for the XP models.

Results are generally consistent with expectations. The positive impact of intangible investments on FDI is borne out by the significant and positive parameter estimates on $\lambda(\gamma' z_i)$ which incorporates the effect of intangibles in the SSM variations. A direct relationship between intensity of FDI shipments and a firm's dominance in its home market appears to be confirmed, with a significantly positive coefficient for SIZE, as hypothesized. The parameter estimates for AMS20 and NINT are also found to be positive, compared to ambiguous priors, thus seemingly strengthening the conclusion that home market dominance leads a firm to invest in foreign affiliates.

As hypothesized, the extent to which firms specialize in food and/or beverage manufacture (SPEC) appears to have no clear impact on FDI (with the exception of modestly significant positive coefficients in Version B), while PD20CT directly effects FDI intensity. The latter is consistent with interpretations of PD20CT as either an indicator of product diversity or product complexity.

Profitability

The results from models estimating the impacts of FDI, XP, and the combined effect of both on firm profitability, measured in terms of net income as a percent of total assets (NITA), are shown in Table 12.4.

There are 312 observations used in the models including FDI as an explanatory variable and 105 observations in the models fit to export data.

TABLE 12.4 Parameter Estimates for a Model of Profitability (NITA)

	Model		
Variable	$f(FDI)$ SSM	$f(XP)$ OLS	$f(FDI,XP)$ SSM
$\lambda(\gamma' z_i)$	-0.0179**		-0.0173**
FDI	0.0421*		0.0427*
XP		-0.0535	0.0506
Adj. R-square	0.177	0.325	0.176
F-value	4.042*	3.786*	3.888*
Sample Size	312	105	312

Note: *, ** significant at the one and five percent levels, respectively.

Because the parameter estimates for λ are statistically significant in the SSM variants that include FDI, those estimates are reported for the models containing FDI. The OLS results are reported for the model with only XP.

Since the relationships fitted to explain NITA using FDI data are affected by the probability that a firm reports financial data on foreign operations we find that:

$$\frac{\partial \text{NITA}}{\partial d^*} > 0$$

indicating that as a firm has a higher tendency to report FDI its net return on assets is more than otherwise.

The estimated models demonstrate significant F-values. However, the adjusted R-square values do not show very good fits, indicating that factors other than extent of international market involvement explain much of inter-firm profit variability. Parameter estimates show a significant and positive effect of FDI and an insignificant effect of XP on NITA.

Conclusions

This study has demonstrated that sensible results regarding determinants of the foreign market behavior of the world's leading food and beverage manufacturing firms can be obtained by using firm-level financial and operating data, even in the presence of incomplete reporting. Heckman's (1979) procedure for adjusting linear regression results to account for the possibility that reporting and non-reporting firms differ in the phenomena of interest appears to be a viable means for dealing with the problem of non-reporting.

Expectations generated from conceptual reasoning and previous empirical studies regarding the impacts of a number of industrial-type factors on firms' export propensities and intensity of foreign direct investment are confirmed for the population of firms with food and/or beverage manufacturing firms included in CIFAR. To summarize, these are:

1. A firm's dominance in its home market as measured by its aggregate share of home-country manufactured food and beverage market is negatively related to exports and positively related to FDI.

2. When home market dominance is expressed as size of firm, smaller firms are more inclined to export; larger firms have greater intensity of FDI.

3. Using net income as a percent of total sales as a proxy for price-cost margin, results are consistent with other measures of home market dominance in explaining FDI but appear to offer little insight regarding export propensity.

4. Investment by a firm in intangibles or firm-specific assets is a direct cause of foreign direct investment; insufficient data prevented testing the impact of intangibles on exports, where expectations are ambiguous.

5. Specialization by a firm in the manufacture of food and/or beverage products directly enhances exports but has no significant impact on FDI.

6. Product diversity or differentiation discourages exports but encourages FDI.

Two conclusions follow from the profitability analysis: (1) for food and beverage manufacturing firms, foreign direct investment is a strategic behavior in the sense that it is profit-enhancing, and (2) the dominance of FDI relative to exports as an international market strategy for these firms corresponds with relative profit opportunities.

Throughout the study, although not explicitly reported here, the estimated coefficients for the dummy variables revealed idiosyncracies in the reporting habits of firms in different industries and in different countries, with weak evidence that more commonality exists among firms in Western European countries than elsewhere.

This study also reveals several needs for subsequent research. The desirability of a common reporting protocol for financial and operating information by firms across all countries is obvious. Such a protocol should not only standardize definitions and classifications, but also expand the scope of reporting on such phenomena as exports by source and destination, foreign operations, through-put and turnover by product line in uniformly-defined geographic markets, expenditures on advertising and research and development, and the value of intangible assets. Future empirical research needs to incorporate other conceptually important determinants of international market behavior such as scale and scope economies, research and development activities other than as embodied in intangibles, and explicit measures of market power in both home and host markets. Future theoretical research is needed to clarify, *inter alia*, why, based on properly motivated models of firm behavior, home market dominance or market power is observed to be negatively related to exports and positively to foreign direct investment. Both industrial and trade policies would be enlightened as a result.

Notes

1. A proxy for home market power.
2. U.S. Standard Industrial Classifications.
3. Specified at the 3-digit SIC level.
4. Net income as a percent of total assets.
5. In the case of the FDI models, the dependent variable can have a value of 0, e.g. no shipments from foreign affiliates. Consequently, OLS results were compared with results using Tobin's (1958) procedure assuming a truncated normal distribution for y. The Tobit results are not reported as they correspond with the OLS results.
6. A confirmation of these results for FDI was obtained from a related model in which the ratio of all foreign sales (exports and FDI) was the dependent variable. See Vörös (1992) for further details.
7. Estimated parameters for the zero-one variables indicate fairly consistent industry idiosyncracies associated with preserved fruits and vegetables and miscellaneous prepared foods, and with all countries except Germany and the UK. As in the XP models, the results are not sensitive to year of observation.

References

Bhagwati, J.N., ed. 1981. *International Trade: Selected Readings*. Cambridge, MA: MIT Press.

Casson. M. 1987. "Multinational Firms," in Clark, R. and T. McGuinnes, eds., *The Economics of the Firm*. Oxford: Basil Blackwell.

Caves, R.E. 1985. "International Trade and Industrial Organization: Problems, Solved and Unsolved." *European Economic Review* 28:377-395.

CIFAR. 1992. *Global Company Handbook*. Princeton, NJ: Center for Financial Analysis and Research.

Connor, J.M. 1983. "Foreign Investment in the U.S. Food Marketing System." *American Journal of Agricultural Economics* 65:395-404.

Ethier, W.J. 1986. "The Multinational Firm." *Quarterly Journal of Economics* 100:807-833.

Fort, R. and W. Hallagan. 1987. "Who Bids the Most for Market Power?" *Economic Inquiry* 25:671-680.

Glejser, H., A. Jacquemin, and J. Petit. 1980. "Exports in an Imperfect Competition Framework: An Analysis of 1,446 Exporters." *Quarterly Journal of Economics* 94:507-524.

Goldberger, A.S. 1964. *Econometric Theory.* New York, NY: Wiley Press.

Greene, W.H. 1991. *LIMDEP Version 6.0.* Bellport, NY: Econometric Software Inc.

Grubaugh, S.C. 1987. "Determinants of Direct Foreign Investment." *Review of Economics and Statistics* 69:149-152.

Handy, C.R. and D.R. Henderson. 1994. "Foreign Direct Investment in Food Manufacturing Industries," in M. Bredahl, P. Abbott, and M. Reed, eds., *Competitiveness in International Food Markets.* Boulder, CO: Westview Press.

Heckman, J.J. 1979. "Sample Selection Bias as a Specification Error." *Econometrica* 47:153-161.

Helpman, E. and P.R. Krugman. 1985. *Market Structure and Foreign Trade.* Cambridge, MA: MIT Press.

Henderson, D.R. and C.R. Handy. 1993. "Globalization of the Food Industry." Paper presented at the conference, *Food and Agricultural Marketing Issues for the 21st Century.* Orlando, FL, January 14-15.

Henderson, D.R. and S.D. Frank. 1990. "Industrial Organization and Export Competitiveness of U.S. Food Manufacturers." NC-194 Occasional Paper OP-4. Columbus, OH: Ohio State University.

Hirschberg, J.G., J.R. Dayton and P.R. Vörös. 1992. "Firm Level Data: A Compendium of International Data Sources for the Food Processing Industries." NC-194 Occasional Paper OP-34. Columbus, OH: Ohio State University.

Hladik, K.J. 1985. *International Joint Ventures.* Lexington, MA: Lexington Books.

Horstmann, I.J. and J.R. Markusen. 1989. "Firm-Specific Assets and the Gains from Direct Foreign Investment." *Economica* 56:41-48.

Jacquemin, A. 1981. "Imperfect Market Structure and International Trade -- Some Recent Research." *Kyklos* 35:75-93.

Jasinowski, J. 1991. "America's Secret Weapon: Trends Show the Power of Manufacturing Exports." *Industry Week* 240:47.

Koo, A.Y.C. and S. Martin. 1984. "Market Structure and U.S. Trade Flows." *International Journal of Industrial Organization* 2:173-197.

Kumar, N. 1987. "Intangible Assets, Internalization and Foreign Production: Direct Investments and Licensing in Indian Manufacturing." *Weltwirtschaftliches Archiv* 123:325-45.

Lipsey, R.E. 1991. "The Competitiveness of the U.S. and of U.S. Firms." Paper presented at the 11th Ministry of Finance-NBER Joint Conference, *The Competitiveness of U.S. Industries and its Implications on U.S.-Japan Relationships in the Future.* Tokyo.

Lyon, B. 1981. "Price-Cost Margins, Market Structure and International Trade," in D. Currie, D. Peel, and W. Peter, eds., *Microeconomic Analysis: Essays in Microeconomics and Economic Development.* London: Croom Helm.

_____. 1989. "An Empirical Investigation of U.K. Manufacturing's Trade With the World and the EEC.: 1968 and 1980," in D.B. Audretsch, L. Sleuwagen, and H. Yamawaki, eds., *The Convergence of International and Domestic Markets.* Amsterdam:North Holland.

Maddala, G.S. 1983. *Limited-Dependent and Qualitative Variables in Econometrics.* New York, NY: Cambridge University Press.

Marvel, H.P. 1992. "Perspectives on Imperfect Competition and International Trade," in I.M. Sheldon and D.R. Henderson, eds., *Industrial Organization and International Trade: Methodological Foundations for International Food and Agricultural Market Research.* North Central Regional Research Publication Number 334. Columbus, OH: The Ohio State University, pp. 5-29.

_____. 1980. "Foreign Trade and Domestic Competition." *Economic Inquiry* 18:103-122.

Pagoulatos, E. 1992. "Empirical Studies of Industrial Organization and Trade: A Selective Survey," in I.M. Sheldon and D.R. Henderson, eds., *Industrial Organization and International Trade: Methodological Foundations for International Food and Agricultural Market Research.* NC-194 Research Monograph No.1. Columbus, OH:Ohio State University.

Pagoulatos, E. and R. Sorensen. 1975. "Domestic Market Structure and International Trade: An Empirical Analysis." *Quarterly Review of Economics and Business* 16:45-59.

Porter, M.D. 1990. *The Competitive Advantage of Nations.* New York, NY: The Free Press.

Rapp, W.V. 1976. "Firm Size and Japan's Export Structure," in H. Patrick, ed., *Japanese Industrialization and its Social Consequences.* Berkeley, CA: University of California Press.

Shaked, A. and J. Sutton. 1990. "Multiproduct Firms and Market Structure." *Rand Journal of Economics* 21:45-62.

_____. 1987. "Product Differentiation and Industrial Structure." *Journal of Industrial Economics* 36:131-146.

Sheldon, I.M. 1992. "Comments on the Interface Between Industrial Organization and International Trade." NC-194 Occasional Paper OP-30. Columbus, OH: Ohio State University.

Sutton, J. 1991. *Sunk Costs and Market Structure: Price Competition, Advertising, and the Evolution of Concentration.* Cambridge, MA: MIT Press.

Tobin, J. 1958. "Estimation of Relationships for Limited Dependent Variables." *Econometrica* 26:24-36.

Vörös, P. 1992. "Industrial Determinants of International Trade and Foreign Investment by Food and Beverage Manufacturing Firms." Unpublished MS Thesis, Department of Agricultural Economics and Rural Sociology, Ohio State University.

White, L.J. 1974. "Industrial Organization and International Trade: Some Theoretical Considerations." *American Economic Review* 64:1013-1020.

Yu, C.J. 1990. "The Experience Effect and Foreign Direct Investment." *Weltwirtschaftlches Archiv* 126:561-580.

13

An Empirical Model for Examining Foreign Direct Investment in the Processed Food Industry

Albert J. Reed

Introduction

Keeping abreast of movements in consumer behavior is important for food manufacturing firms undertaking foreign direct investment (Handy and Seigle 1988). Handy and Henderson (1992) note the most 'profitable' multinational food manufacturing firms with foreign affiliates value intangible assets highly. Intangible assets include data gathering information systems (Connor 1983).

This chapter maintains that foreign direct investment (FDI) enables a firm to change its response to consumers in a foreign market. Compared to an export strategy, this chapter maintains FDI imparts three advantages to firms. First, the firm may exploit a comparative advantage in management skill when located within the market (Handy and MacDonald 1989). Management facilitates internal adjustment when the firm adjusts supply in the short run. Second, in large markets, the firm may realize economies of scale or size. In this chapter, economies of scale or size alter the shape of the long-run cost function. Third, if the firm locates within the foreign market, it acquires more precise information on consumer behavior. The chapter infers the importance of each of the three advantages based on the firm's change in response to consumers.

At the heart of this inference is an economic model with a structural parameter assigned to each of the three advantages. The model's solution formally links each of the three parameters to a response coefficient associated with consumer demand shift data. When a firm changes from an export strategy to FDI, the study maintains that each of the three structural parameters change. The empirical task is to compare the change in the response coefficient due to a change in each of the three structural parameters. A coefficient that is highly sensitive to changes in the parameter describing data accuracy, for example, suggests the firm's quest for better information motivated it to undertake FDI.

The theory of information is made precise in this study with the assumption that firms form rational expectations in the sense of Sargent (1978), Lucas (1979), or Townsend (1983). That is, firms form expectations consistent with the solution to an explicit, dynamic optimization problem. In the work of Sargent (1978), agents possess full information, so forecast errors are unpredictable. However, the present work is more closely akin to the works of either Lucas (1979) or Townsend (1983), because firms possess less-than-full information sets. The consequence is serially correlated forecast errors that add persistence to economic time-series data. The concept of rational agents with less-than-full information provides a useful description of firms responding to consumers in a distant market.

Both the signal extraction and the bootstrap procedures used below are potentially valuable tools for students of foreign direct investment. Foreign affiliate data usually consists of a short time-series, and usually contain observation error. In this study, the signal extraction technique permits parameter estimation when equations contain a variable observed with error. The bootstrap is used to compute the distribution of parameter estimates. The bootstrap can also be used to investigate how the distribution of parameter estimates change as a short time-series is lengthened (Freedman 1981).

The Economic Model

This section formalizes the ideas introduced in the previous section. The first part of the model captures consumer behavior. The key parameter in the first part of the model is the variance of the observation error in data on shifts in consumer demand. The second part of the model focuses on the firm's technology. The key parameters in the second part of the model are long-run and short-run cost parameters. When firms undertake FDI, this study maintains firms observe more reliable data on consumer demand shifts, and they realize long-run and short-run economies. The solution of the economic model links the firm's three structural parameters to its response to demand shift data.

Consider consumers of product q located in a foreign market. Foreign consumers pay price p_t, consume quantity q_t, and consume quantities of other related goods. q-producers, located either exclusively at home or within the foreign market, accurately observe data on own-price, p_t, and own-quantity, q_t, but observe data on the consumption of other goods with error. From this description, Appendix A justifies the inverse consumer demand function for q as:

$$P_t = -A_1 q_t + z_t \qquad (1)$$

where A_1 is a slope parameter and z_t is a weighted sum of consumption data of other goods. z_t denotes the composite demand shift data observed by firms. θ_t represents the actual consumption of other goods consumed. θ_t is not observed by firms, but as Appendix A points out, it is related to the observed data by:

$$z_t = \theta_t + u_t \qquad (2)$$

in which u_t is a composite observation error, consisting of a weighted sum of individual observation errors of each of the other (than q) goods consumed. Formally, let E represent the mathematical expectations operator, and $Eu_t = 0$, $Eu_t u_{t-s} = 0$ for $s \neq 0$, and $Eu_t^2 = \sigma_u^2$. Furthermore, firms know the actual, unobserved demand shifter follows:

$$\theta_{t+1} = \rho\theta_t + \varepsilon_{t+1} \qquad (3)$$

where $E\varepsilon_t = 0$, $E\varepsilon_t \varepsilon_{t-s} = 0$ for $s \neq 0$, $E\varepsilon_t^2 = \sigma_\varepsilon^2$, and $Eu_t\varepsilon_s = 0$ for all s and t. ε_t represents the fundamental innovation (i.e., ε_t is white noise and orthogonal to past θ) associated with θ_t. The first part of the model permits an analysis of a firm's response when it obtains more reliable information on demand shifts. Based on information available to it each period, the firm chooses the quantity of the variable input, f_t, to hire in the production of food, q_t. For simplicity, the firm is assumed to produce q_t according to $q_t = \alpha f_t$, where α is an output-input coefficient. Furthermore, q_t is a differentiated product, a feature that bestows monopoly power on the firm. The firm's return function is:

$$\pi_t = -A_1\alpha^2 f_t^2 + \theta_t\alpha f_t - \left(\frac{h}{2}\right)f_t^2 - \left(\frac{d}{2}\right)(f_t - f_{t-1})^2$$

and the firm attempts to:

$$\max_{\{f\}} \ E \sum_{t=0}^{\infty}\beta^t\pi_t|\Omega_o. \qquad (4)$$

Here $\beta = 1/(1+r)$ and r is the interest rate. Also, the information available to firms at time t is summarized in the set Ω_t, defined as:

$\Omega_t = \{ \, E\,(\theta_t|\Omega_t), E\,(\theta_{t-1}|\Omega_{t-1}), ..., P_t, P_{t-1}, ..., f_{t-1}, f_{t-2}, ... \}.$

Ω_t is a less-than-full information set. It is less-than-full because it does not contain actual shifts in consumer demand. Information regarding demand shifts is embodied in the firm's present and past expectations of consumer demand shifts.

The literature suggests two types of costs may be reduced when a firm undertakes FDI. Handy and Henderson (1992) claim the size of the firm is correlated with the firm's 'foreign presence'. The term $(h/2)f_t^2$ is a long-run economies of scale or size term, with constant returns when h = 0. (Townsend 1983) When firms undertake FDI, this study maintains firms realize some long-run economies, and reduce their h parameters. Reducing a positive h parameter (with d > 0) translates into a larger response to shifts in consumer demand (Reed 1992). Firms undertaking FDI may also have a comparative advantage because of superior management skill (Handy and MacDonald 1989). The term $(d/2)(f_t - f_{t-1})^2$ maps changes in the factor of production, $(f_t - f_{t-1})$, to a management requirement $(d/2)(f_t - f_{t-1})^2$. 'Small' d parameters are associated with 'high' management skill. Reducing a positive d parameter (with h > 0) translates into a larger response to shifts in consumer demand (Reed 1992). The cost function embodied in the above return function formally describes the two types of costs that may be reduced when firms undertake FDI.

A change in the h or d parameter alters a firm's response to an expected consumer demand shift. To see this, note the solution to the monopoly problem described in equations (1)-(4) is:

$$f_t = \lambda f_{t-1} + \left(\frac{\alpha\lambda}{d[1-\lambda\beta\rho]} \right) E\,(\theta_t|\Omega_t) \tag{5}$$

(Reed 1992), where λ satisfies:

$$\beta\lambda^2 - \left(\frac{2A_1\alpha^2 + h + d(1+\beta)}{d} \right)\lambda + 1 = 0.$$

Hence, the cost-of-adjustment (d) and scale economies (h) parameters define λ, and therefore define the firm's response to expected demand shifts.

A change in the variance of the observation error, σ_u^2, alters a firm's response to observed demand shifts. To see this, one needs a formal statement of how a rational firm forms expectations on the unobserved, actual shift in demand. The Kalman filter provides such a statement (e.g., Sargent 1987). The equations of the Kalman filter can be written as:

$$E\left(\theta_t | \Omega_t\right) = (1 - K_t)E\left(\theta_t | \Omega_{t-1}\right) + K_t z_t, \tag{6}$$

and:

$$P_{t+1|t} = (\rho - \rho K_t)^2 P_{t|t-1} + \sigma_\varepsilon^2 + \rho^2 K_t^2 \sigma_u^2 \tag{7}$$

where:

$$P_{t|t-1} \equiv E\left\{\theta_t - E\left(\theta_t | \Omega_{t-1}\right)\right\}^2,$$

and:

$$K_t = P_{t|t-1}(P_{t|t-1} + \sigma_u^2)^{-1}. \tag{8}$$

Equations (6) and (7) define the conditional mean and variance of the unobserved demand shift variable. Equation (6) indicates the current period expectation or forecast of θ_t is a weighted sum of last period's forecast of θ_t and current period demand shift data, z_t. The weight parameter, K_t, is termed the Kalman gain. According to equations (7) and (8), the Kalman gain changes over time and varies inversely with the variance of the noise in the data. The relationship between the Kalman gain and the variance of the observation error in demand shift data offers two economic interpretations regarding FDI.

First, equations (6) and (8) reveal how firms learn from mistakes. Notice equation (6) can be rewritten as:

$$E\left(\theta_t | \Omega_t\right) = E\left(\theta_t | \Omega_{t-1}\right) + K_t(z_t - E\left(z_t | \Omega_{t-1}\right)). \tag{6'}$$

In current period t, firms see a new observation on consumer demand shifts, z_t. The new observation reveals a new data forecast error or mistake, $z_t - E\left(z_t | \Omega_{t-1}\right)$. Equation (6') indicates firms use this mistake to revise their current period forecast of actual, unobserved shifts in demand (θ_t). The weight firms give to the data forecast error is termed the Kalman gain. Equation (8) indicates the Kalman gain, K_t, is inversely related to the variance of the noise in the data. This is intuitive. Rational firms realize forecast errors of noisy data may be largely due to noise rather than to movements in actual demand shifts. Hence, rational firms will not significantly revise their current period forecasts of actual demand shifts. On the other hand, when data contain little noise, a data forecast error more likely signals a change in the actual demand shift. Rational firms place significant importance on forecast mistakes when the data contain a relatively small amount of noise.

Second, equations (6) and (8) reveal how firms respond to newly available data. Equations (6) and (8) indicate the noisier the data, the less weight firms place on new data in formulating forecasts. The reason is simple: the noisier the data, the less reliable is the data in tracking unobserved demand shifts.

Equation (8) also reveals that improvements in the reliablility of the data do not always lead firms to alter their reponse. When the variance of the underlying demand shift, $P_{t|t-1}$, dominates the observation error variance, σ_u^2, reducing σ_u^2 by undertaking FDI will not significantly alter a firm's response. Volatile changes in consumer behavior in transitional economies such as Eastern Europe may outweigh the observation error variance to the extent improvements in data quality will not justify FDI.

To see how changes in the three structural parameters of the model alter a firm's response to observed demand shifts, equations (5) and (6) imply:

$$f_t = \lambda f_{t-1} + \left(\frac{\alpha\lambda}{d[1-\lambda\beta\rho]}\right)\{(1-K_t)E\,(\theta_t|\Omega_{t-1}) + K_t z_t\}. \tag{9}$$

The coefficient associated with demand shift data, z_t,

$$\left(\frac{\alpha\lambda}{d[1-\lambda\beta\rho]}\right)K_t,$$

increases as σ_u^2 is reduced, and is sensitive to changes in the h and d parameters. The empirical question is how sensitive is the above coefficient to changes in each of the three structural parameters?

This study computes long-run or steady-state sensitivity elasticities of response to reductions in the h, d, or σ_u^2 parameters. Ignoring the E $(\theta_t|\Omega_{t-1})$ term in equation (9), so equation (9) can be inverted to:

$$f_t = (1 - \lambda L)^{-1}H_t z_t,$$

or,

$$f_t = (H_t + \lambda H_{t-1}L + \lambda^2 H_{t-2}L^2 + ...)z_t,$$

where,

$$H_t = \left(\frac{\alpha\lambda}{d[1-\lambda\beta\rho]}\right)K_t.$$

Analysis of the steady-state behavior of the Kalman filter equations (7) and (8) indicates that for any value of ρ, and any starting value, P_0, P_t and K_t converge to the constants P and K as $t \rightarrow \infty$. Specifically, $P = \lim_{t \rightarrow \infty} P_t$ and $K = \lim_{t \rightarrow \infty} K_t$ (Sargent 1987). Steady-state sensitivity elasticities are computed by letting $z_t = z_{t-1} = ... = z$, so the partial derivative is:

$$\frac{\partial f_t}{\partial z} = H(1 + \lambda + \lambda^2 + ...) \equiv \Delta$$

where:

$$H = \left(\frac{\alpha\lambda}{d[1-\lambda\beta\rho]} \right) K.$$

The idea is to compute:

$$\frac{\partial}{\partial h}\left(\frac{\partial f_t}{\partial z}\right) \times \left(\frac{h}{\Delta}\right), \quad \frac{\partial}{\partial d}\left(\frac{\partial f_t}{\partial z}\right) \times \left(\frac{d}{\Delta}\right), \quad \frac{\partial}{\partial \sigma_u^2}\left(\frac{\partial f_t}{\partial z}\right) \times \left(\frac{\sigma_u^2}{\Delta}\right)$$

which are the scale-economy-response elasticity, the adjustment-cost-response elasticity and information-response elasticity. The above three terms express the percentage change in the response coefficient to demand shift data due to a one-percent decrease in the h parameter, a one-percent decrease in the d parameter, and a one-percent decrease in the σ_u^2 parameter, respectively.

The model also reveals the persistence added to an economic time-series because firms possess less-than-reliable data. A time-series persistently deviates from its long-run or steady-state equilibrium following a market surprise or shock. A time-series with 'high' persistence takes a long time to return to its steady state path. Townsend (1983) illustrates how the added uncertainty of less-than-reliable data adds persistence to an economic time-series. The implication is that a shock to demand shifts may induce a less-than-smooth supply response. To the extent FDI permits firms to observe more reliable data, FDI reduces forecast error persistence and smoothes a supply response.

The degree of added persistence is quantified. Note, equation (3) implies:

$$\theta_t = \rho\theta_{t-1} + \varepsilon_t.$$

The rational expectation of the underlying demand shift is:

$$E\left(\theta_t | \Omega_t\right) = (1 - K_t)\rho E\left(\theta_{t-1} | \Omega_{t-1}\right) + K_t \rho \theta_{t-1} + K_t \varepsilon_t + K_t u_t$$

from which it follows:

$$\left[\theta_t - E\left(\theta_t | \Omega_t\right)\right] = (\rho - K_t \rho)\left[\theta_{t-1} - E\left(\theta_{t-1} | \Omega_{t-1}\right)\right] - K_t \varepsilon_t - K_t u_t.$$

Hence, unless $K_t = 1$, forecast errors display persistence because they are serially correlated. Serially correlated forecast errors obtain despite the assumption of rational expectations. According to equation (8), if $K_t = 1$, $\sigma_u^2 = 0$, $|\rho - \rho K_t| = 0$, and no persistence is added to the time series. The coefficient $|\rho - \rho K|$ measures the degree of persistence added to a data series because agents possess less-than-full information. A large value of the coefficient denotes a series with high persistence added to the time series, and a small value denotes a series with low persistence added to the time series.

Statistical Model

The previous section illustrated how dynamic economic theory directs firms to form a rational expectation of an unobserved variable. The Kalman filter was central to the analysis. The present section focuses on representing an estimable econometric model. The Kalman filter is again central to the discussion.

The works of Burmeister and Wall (1987) and Stoffer and Wall (1991) combine the Kalman filter with the bootstrap. The bootstrap is important not only because it permits the computation of standard errors of parameter estimates, but also because it permits the computation of standard errors of the sensitivity elasticities. The methodology presented in this section closely follows the works of Burmeister and Wall (1987) and Stoffer and Wall (1991). The main difference is the present work uses a two-step rather than a one-step estimation procedure.

Typically, observable economic time-series define observable innovations (i.e., white noise with orthogonality conditions) which are used to compute the model's parameter estimates. However, the above model includes the expectation of an unobservable demand shift variable. The Kalman filter uses the conditional expectation to define observable innovations:

$$\hat{\varepsilon}_{t|t-1} \equiv z_t - E\left(z_t | \Omega_{t-1}\right) = z_t - E\left(\theta_t | \Omega_{t-1}\right).$$

By recursive expectations,

$$E\left(\theta_{t+1} | \Omega_t\right) = \rho E\left(\theta_t | \Omega_{t-1}\right) + \rho K_t \mathcal{E}_{t|t-1}.$$

In this form, observable innovations define an observable expectation of the unobserved demand shift variable.

The recursive nature of the Kalman filter is easily seen by rewriting the innovations form of the model in the state-space form:

$$\begin{bmatrix} E\left(\theta_{t+1} | \Omega_t\right) \\ z_t \end{bmatrix} = \begin{bmatrix} \rho & 0 \\ 1 & 0 \end{bmatrix} \begin{bmatrix} E\left(\theta_t | \Omega_{t-1}\right) \\ z_{t-1} \end{bmatrix} + \begin{bmatrix} \rho K_t \\ 1 \end{bmatrix} \mathcal{E}_{t|t-1} \tag{10}$$

with

$$\begin{aligned} \Sigma_{t|t-1} &\equiv E\, \mathcal{E}_{t|t-1}^2 \\ &= E\left(z_t - E\left(\theta_t | \Omega_{t-1}\right)\right)^2 = E(\theta_t - E\left(\theta_t | \Omega_{t-1}\right) + u_t)^2 = P_{t|t-1} + \sigma_u^2 \end{aligned} \tag{11}$$

$$P_{t+1|t} = (\rho - \rho K_t)^2 P_{t|t-1} + \sigma_\varepsilon^2 + \rho^2 K_t^2 \sigma_u^2 \tag{12}$$

$$K_t = P_{t|t-1} \Sigma_{t|t-1}^{-1}. \tag{13}$$

The first entry of the state vector of equation (10) defines the conditional expectation of the unobserved demand shifter, and is used in the second step of the estimation procedure. Equations (11) to (13) are equivalent to equations (6) to (8). Equations (11) to (13) express the Kalman gain in terms of the covariance of the model's innovations. In turn, the Kalman gain defines the recursive covariance term.

The first step of the estimation procedure computes the value of the vector $\delta_1 = [\rho, \sigma_\varepsilon, \sigma_u]'$ that minimizes the negative of the Gaussian likelihood i.e.,

$$\delta_1 = [\rho, \delta_\varepsilon, \delta_u] = \text{argmin } \mathcal{L}_T(\delta_1)$$

in which,

$$\mathcal{L}_T(\delta_1) = \mathcal{L}_T(\rho, \sigma_\varepsilon^2, \sigma_u^2) = -\Sigma_{t=1}^T \{\ln |\Sigma_{t|t-1}| + \mathcal{E}_{t|t-1}' \Sigma_{t|t-1}^{-1} \mathcal{E}_{t|t-1}\}$$

and T represents the size of the sample. The first step of the estimation procedure computes a Gaussian Maximum Likelihood (GML) estimate of the δ_1 vector.

The second step of the estimation procedure uses the reduced-form equation (9), first-step estimates, and predetermined parameter values to compute estimates of the technology vector $\delta_2 = [\lambda, d]'$. An estimate of the scale economies parameter, h, is then uncovered. Using first-step estimates of ρ and $E(\theta_{t+1}|\Omega_t)$, the second step computes the value of δ_2 that minimizes:

$$\delta_2 = [\hat{\lambda},\hat{d}]' \equiv \text{argmin } s_T(\delta_2|\beta,\rho,E(\theta_t|\Omega_{t-1}))$$

in which,

$$s_T (\delta_2|\beta,\rho,\hat{E}(\theta_t|\Omega_{t-1})) = \frac{1}{T}\Sigma_{t=1}^T e_t^2 =$$

$$\frac{1}{T} \Sigma_{t=1}^T [f_t - \lambda f_{t-1} - \left(\frac{\alpha\lambda}{d[1-\lambda\beta\rho]}\right)\{(1-K_t)\hat{E}(\theta_t|\Omega_{t-1}) + K_t z_t\}]^2$$

where '^' denotes the first-step estimate. The values of α and β are predetermined. Then, the estimate of h is obtained by evaluating:

$$\beta\lambda^2 - \left(\frac{2A_1\alpha^2 + h + d(1+\beta)}{d}\right)\lambda + 1 = 0, \tag{14}$$

at the estimated values of λ and d, and the predetermined values of α, β, and A_1, Hence, the second step of the estimation procedure computes least squares (LS) estimates of the δ_2 vector and the h parameter.

Inference on GML or LS estimates is straightforward when samples are large or are normally distributed. However, when samples are neither large nor normal, the researcher is left with little guide to inference. The bootstrap facilitates inference by computing the small sample distribution of parameter estimates.

Freedman (1981) warns that if the bootstrap is to provide unbiased estimates, the model's error terms must be independently distributed with a common distribution. However, equations (11) to (13) indicate the model's innovations are not independently distributed, as the covariance term is not constant over the sample. Freedman (1981) further warns that sampling from residuals that do not sum to zero results in biased estimates. The state-space vector equations (10) and the reduced-form equation (9) lack constant terms, so the residuals need not sum to zero. However, the transformed residuals:

$$\xi_{t|t-1} = \Sigma_{t|t-1}^{-1/2} (\varepsilon_{t|t-1} - \bar{\varepsilon})$$

$$\mu_t = \sigma_e^{-1}(e_t - \bar{e}),$$

(where the bar denotes sample means) sum to zero and have at least the first two moments of their distribution in common. Bootstrap draws from the transformed residuals should result in unbiased estimates.

The bootstrap is performed by computing parameter estimates using the two-step procedure, and collecting and transforming the residuals. The transformed residuals are then drawn with replacment. Letting '^' denote the original parameter estimate, bootstrap samples of the data are generated according to:

$$\begin{bmatrix} E(\theta_{t+1}|\Omega_t)^* \\ z_t^* \end{bmatrix} = \begin{bmatrix} \rho & 0 \\ 1 & 0 \end{bmatrix} \begin{bmatrix} E(\theta_t|\Omega_{t-1})^* \\ z_{t-1}^* \end{bmatrix} + \begin{bmatrix} \rho\hat{K}_t\hat{\Sigma}_{t|t-1}^{1/2} \\ \hat{\Sigma}_{t|t-1}^{1/2} \end{bmatrix}\xi_{t|t-1}^* + \begin{bmatrix} \rho\hat{K}_t \\ 1 \end{bmatrix}\bar{\varepsilon}, \qquad (15)$$

and,

$$f_t^* = \lambda f_{t-1}^* + \left(\frac{\alpha\hat{\lambda}}{d[1-\hat{\lambda}\beta\rho]}\right)\{(1-\hat{K}_t)\,\hat{E}(\theta_t|\Omega_{t-1})^* + \hat{K}_t z_t^*\} + \sigma_e\mu_t^* + \bar{e}.$$

where '*' denotes the bootstrap sample. The two-step estimation procedure is then repeated using bootstrap samples. For B bootstrap samples, there are B bootstrap estimates of δ_1, δ_2, and h. The mean and the standard deviation of the bootstrap estimates are computed in the usual way.

Results

The empirical results presented in this section illustrate how the model is used to interpret secondary data of four food manufacturing industries. The results attempt to quantify the importance of scale economies, management skill, and better information to food firms that have undertaken FDI.

Secondary data for domestic cereal, nonalcoholic beverage, canned fruits and vegetables, and fats and oils industries are analyzed in this section. Table 13.2 provides examples of the firms that have undertaken foreign direct investment within each of the four industries. Seven of the eight firms listed in Table 13.2 had sales exceeding $1.6 billion in 1990, placing them among the top 10 U.S. food manufacturing firms with foreign affiliates (Handy and Henderson 1992).

Conceptually, consumption and retail price data from target economies -- economies in which U.S. food manufacturing firms have invested -- should be used to implement the model. However, such data were not readily available, and U.S. data were used. The most obvious problem with using U.S. demand data is the noise component of the data may not be relevant to U.S. investors.

A time-series on manhours per week forms the single variable input series in the study. The manhours series, $\{f_t\}$, is constructed from the Bureau of Labor Statistics (BLS) monthly data on employment and average weekly hours. The product of the two monthly variables is averaged over three months to form the quarterly manhours per week time series. Quarterly consumer prices, p_t, for each industry are three month averages of the BLS monthly Consumer Price Index (CPI) on selected goods. The means of the sample time-series are reported in Table 13.1.

Predetermined parameter values for the discount factor (β), the output-input coefficient (α), and the slope of the inverse demand function (A_1) are specified for each industry. β is set to 0.95, a value that corresponds to an interest rate of 0.053, and α is set to unity for each of the four industries. Huang's (1988) estimated price flexibilities for cereal, beverages, processed fruits and vegetables, and fats are reported in Table 13.2. Each flexibility is transformed into a linear slope coefficient, A_1, by multiplying the flexibility by the ratio of the sample mean of manhours

TABLE 13.1 Data for Four Food Manufacturing Industries

BLS Labor Series	BLS CPI Series	Manhours (f)	CPI (p) 82-84=100	Sample Interval
Cereal Breakfast Food	Cereal & Cereal Products	811.634	109.969	1978I-92IV
Bottled & Canned Soft Drinks	Carbonated Drinks	4887.045	99.7606	1978I-92IV
Cannd Fruits & Vegetables	Prosecced Fruits & Veg	3738.194	91.172	1972I-92IV
Fats & Oils	Fats & Oils	493.512	89.443	1972I-92IV

Source: Bureau of Labor Statistics.

TABLE 13.2 Example Firms and Estimates of Demand Parameters

	Firms with Foreign Affiliates	Huang's Flexibility Estimate[*]	A_1
Cereal Breakfast Food	Kelloggs Quaker Oats	-0.405	2.989
Bottled & Canned Soft Drinks	Pepsi Coca-Cola	-1.428	69.955
Canned Fruits & Veg	Gerber Ph. Morris/Kraft Campbell Soup	-1.030	42.232
Fats & Oils	Ph. Morris/Kraft CPC International Quaker Oats	-1.157	6.384

Note: [*] Huang (1988).

to the sample mean of the retail price. Huang's (1988) flexibility estimate and the computed value for A_1 are presented in Table 13.2.

Econometric estimation of the parameters of the model requires a time-series for both manhours and observed demand shifts. The time-series of observed demand shifts, $\{z_t\}$, is computed according to equation (1) using the values of A_1 and α, and the sample sequences of retail price and manhours. The sample means are removed from the demand shift and manhours series, so the data represent deviations from the mean. The model is specified without constants, and starting values for the parameters, and the initialization of the conditional covariance term of the Kalman filter is described in detail in Appendix B.

The parameter estimates and their standard errors (in parentheses) are computed from the original, deviation-from-the-mean sample series, and are reported in Table 13.3. The residuals of the two-step estimation procedure are collected and 100 bootstrap samples of the standardized residuals are used to generate 100 bootstrap samples of manhours and demand shifts. The bootstrap data samples consist of the same number of observations as the original sample. Table 13.4 presents the means and standard errors of the 100 parameter estimates.

The bootstrap point estimate is the mean of the bootstrap distribution. The results reported in Table 13.4 indicate the bootstrap point estimates appear to be converging to the point estimates computed using the original data and reported in Table 13.3. Because it is a nonlinear function

TABLE 13.3 Parameter Estimates, Original Data*

	ρ	σ_ε	σ_u	λ	d	h
Cereal	0.91	23.09	24.48	0.80	7.78	-5.53
	(.38)	(29.3)	(21.1)	(.06)	(4.0)	
Soft	1.00	7639.11	14305.04	-0.25	10.73	-205.79
Drink	(.15)	(11280.6)	(9272.0)	(0.13)	(3.3)	
Canned	0.91	278.76	1431.91	-0.18	-0.38	-81.57
Fruits &	(.38)	(594.3)	(544.3)	(0.11)	(.19)	
Veg						
Fats &	0.91	99.69	34.65	-0.03	0.21	-19.48
Oils	(.38)	(66.9)	(105.2)	(.04)	(.22)	

Note: * Based on 100 bootstrap samples. Standard errors are given in parentheses.

of other parameter estimates, the standard error of the point estimate of h is difficult to compute using only the original data and the two-step prodcedure. The bootstrap standard error, however, is computed in the same way for all parameter estimates: it is the standard deviation of the bootstrap point estimates. The results presented in Table 13.4 suggest the estimated standard errors reported in Table 13.3 and based on large sample theory overstate the estimates computed from a small sample.

The most striking feature of the results presented in Tables 13.3 and 13.4 is the negative point estimate of the h parameters for each of the four industries. A negative h parameter indicates long-run increasing returns to scale. Furthermore, the relatively small standard errors presented in Table 13.4 suggest the h parameter is precisely estimated. The results presented in Tables 13.3 and 13.4 suggest domestic firms that have undertaken FDI have realized long-run scale economies.

The consequence of negative values for h are reflected in the negative signs on some of the sensitivity elasticity estimates reported in Table 13.5. Table 13.5 reports the percent change in the response coefficient associated with demand shift data when firms realize a one-percent decrease in the h parameter, a one percent decrease in the d parameter, or a one percent decrease in the variance of the noise in the data. The negative values reported in Table 13.5 suggest as firms undertake FDI, they become less responsive to observed demand shifts. The negative signs owe partly to the long-run increasing returns to scale estimate.

Previous work imposes curvature conditions on a similar economic model (Reed 1992). In particular, the curvature conditions impose positive h and d parameters on the economic model, and a comparative

TABLE 13.4 Parameter Estimates, Bootstrap Data[*]

	ρ	σ_e	σ_u	λ	d	h
Cereal	0.854	23.269	20.469	0.601	1.548	-5.609
	(.13)	(10.1)	(10.2)	(.10)	(1.1)	(.10)
Soft Drink	0.988	7062.96	14151.11	-0.406	12.906	-203.2
	(.05)	(3612.8)	(1931.2)	(.10)	(1.7)	(4.0)
Canned Fruits & Veg	0.881	332.33	1411.98	-0.266	-0.306	-82.55
	(.31)	(420.5)	(315.2)	(0.09)	(.11)	(.66)
Fats & Oils	0.955	104.07	24.018	-0.042	0.173	-19.40
	(.04)	(18.0)	(25.13)	(.09)	(.49)	(.13)

Note: [*] Based on 100 bootstrap samples. Standard errors are given in parentheses.

dynamic analysis implies reductions in the h or d parameters unambiguously increase the responsiveness of firms. Negative estimates of h violate the curvature conditions, and do not impose a sign on the sensitivity elasticities reported in Table 13.5.

The estimated negative sign on h should be interpreted cautiously. While a domestic firm may have achieved long-run scale economies in the domestic market, it is plausible a newly-located plant in a distant market may take a considerable period of time to achieve scale economies. The estimation of a learning-by-doing model may capture the feature that marginal cost functions change over time as firms establish themselves in the market (Reed 1992).

TABLE 13.5 Percent Change in Response Coefficient Due to a 1 Percent Decrease in the h, d, and σ^2_u Parameters[*]

	Scale Economies (h)	Adjustment Cost (d)	Noise Variance (σ^2_u)
Cereal	-15.04	-0.093	+0.298
	(5.2)	(.07)	9.120
Soft Drink	+3.241	-0.005	+0.372
	(.33)	(.07)	(.05)
Canned Fruits & Veg	-82.65	-0.044	+0.492
	(168.5)	(.12)	(.14)
Fats & Oils	+2.926	-0.0003	+0.083
	(.04)	(.004)	(.080)

Note: [*] Based on 100 bootstrap samples. Standard errors are given in parentheses.

Nevertheless, the results of Table 13.5 suggest firms in the cereals, soft drink, and fats and oils industries that have undertaken FDI are sensitive to scale economies. In the case of cereal firms, increased scale economies tend to dampen responsiveness. In the case of soft drink and fats and oils firms, scale economy gains tend to increase responsiveness. The relatively large estimate of the bootstrap standard error associated with the canned fruit and vegetable firms suggest the sensitivity elasticity is imprecisely estimated. The results suggest scale economies provide a strong incentive to cereals, soft drink, and fats and oils manufacturing firms considering FDI.

The results of Table 13.5 also indicate more reliable data on consumer demand shifts leads to positive increases in responsiveness. Although not as large as the change in the responsiveness attributable to scale economies, more reliable information may be at least as important to firms. The reason is that a plant located in a new market may be able to gather more reliable data on consumer shifts more quickly than it can realize long-run scale economies.

The results of Table 13.5 indicate reducing cost-of-adjustment by undertaking FDI does not signifcantly alter a firm's responsiveness. This result suggests the comparative advantage of management skill is an overstated advantage of FDI.

Finally, the results presented in Table 13.6 suggest the persistence stemming from less-than-reliable demand shift data is highest for soft drink and canned fruits and vegetables. The results imply FDI in the soft drink and canned fruits and vegetable industries serves to smooth the supply response following demand shocks in distant markets.

TABLE 13.6 Estimate of $\rho - \rho K^*$

Cereal	Soft Drink	Canned Fruit & Veg	Fats & Oils
0.355	0.588	0.712	0.096
(.18)	(.11)	(.28)	(.10)

Note: * Based on 100 bootstrap samples. Standard errors are given in parentheses.

Conclusions

This chapter began with an obvious conclusion on economic efficiency: if FDI represents a viable alternative to trade, the export response of food manufacturing firms will differ from the response of plants located within the market. Based on the change in responsiveness, this chapter attempts to empirically determine the reasons U.S. food manufacturing firms pursue FDI.

Perhaps it is important to emphasize how the empirical results of the study should not be interpreted. Unlike the work of Ethier (1986), the above framework does not endogenize foreign direct investment. The model is not designed to predict the level of FDI within different food industries. However, the model explicitly recognizes Ethier's comment: "...the internalization of international transactions must be preferable to the arms' length use of markets." Reductions in internal cost-of-adjustment, perhaps due to superior management skill, and realization of internal scale economies alter the plant's marginal cost function. The empirical results do suggest that having undertaken FDI, firms will alter their response more because of scale economy changes than because of internal adjustment cost, or management skill improvements.

The study also does not appeal to the public good nature of information. Handy and MacDonald (1989) argue that the failure of information markets to appropriately allocate the rents to an innovation is corrected by FDI. It is clear that a test of this hypothesis was not and cannot be performed until the concept of information is made precise. The above framework offers a precise and empirically useful definition of information: as domestic firms locate closer to target markets, consumer data contain less noise or observation error. The empirical results suggest firms respond positively to improved information, and the positive responses are statistically significant. From a purely statistical point of view, better information results in a smoother time-series. From the firm's point of view, the gain in information from undertaking an FDI strategy results in a smoother supply response than if the firm exclusively pursues an export strategy.

References

Burmeister, E. and K.D. Wall. 1987. "Unobserved Rational Expectations and the German Hyperinflation with Endogenous Money Supply." *International Economic Review* 28:15-32.

Connor, J.M. 1983. "Foreign Investment in the U.S. Food Marketing System." *American Journal of Agricultural Economics* 65:395-404.

Ethier, W.J. 1986. "The Multinational Firm." *Quarterly Journal of Economics* 51:805-43.

Freedman, D.A. 1981. "Bootstrapping Regression Models." *The Annals of Statistics* 9:1218-1228.

Handy, C.R. and J.M. MacDonald. 1989. "Multinational Structures and Strategies of U.S. Food Firms." *American Journal of Agricultural Economics* 71:1246-54.

Handy, C.R. and D.R. Henderson. 1992. "Foreign Investment in Food Manufacturing." NC-194 Occasional Paper OP-41. Columbus, OH: Ohio State University.

Handy, C.R. and N. Seigle. 1988. "International Profile of U.S. Food Processors." Paper prepared for the Food Distribution Research Society annual meeting, Houston, TX.

Huang, K.S. 1988. "An Inverse Demand System for U.S. Composite Foods." *American Journal of Agricultural Economics* 70:902-09.

Lucas, R.E., Jr. 1979. "An Equilibrium Model of the Business Cycle." *Journal of Political Economy* 83:1113-1144.

Newbold, P. and T. Bos. 1985. "Stochastic Parameter Regression Models." Sage University Paper Series: Quantitative Applications in the Social Sciences, J.L. Sullivan and R.G. Niemi, eds., Beverly Hills, California.

Reed, A.J. 1992. "A Framework for Examining Direct Foreign Investment in the Processed Food Industry." NC-194 Occasional Paper OP-36. Columbus, OH: Ohio State University.

Sargent, T.J. 1978. "Estimation of Dynamic Labor Demand Schedules Under Rational Expectations." *Journal of Political Economy* 86:1009-1044.

Stoffer, D.S. and K.D. Wall. 1991. "Bootstrapping State-Space Models: Gaussian Maximum Likelihood Estimation and the Kalman Filter." *The Journal of the American Statistical Association* 86:1024-1033.

Townsend, R.M. 1983. "Forecasting the Forecasts of Others." *Journal of Political Economy* 91:546-587.

Appendix A. Observation Error on the Demand Shifter

This appendix offers an explanation as to how observation errors enter demand shift data observed by firms by deriving:

$$P_t = -A_1 q_t + z_t \tag{i}$$

$$z_t = \theta_t + u_t \tag{ii}$$

which represent equations (1) and (2) in the text. Consumers in the distant market consume good q and an n-vector, $X_t = [x_{1t}\ x_{2t}...,x_{nt}]'$ of related goods. Suppose consumers obey the inverse demand function:

$$P_t = -A_1 q_t + \beta' X_t + v_t \tag{iii}$$

where $\beta = [\beta_1, \beta_2,...,\beta_n]'$, and where v_t is a purely transitory shock in which $Ev_t = 0$, $Ev_t = \sigma^2_v$, and $Ev_t v_s = 0$ for all $s \neq t$. A_1 and β approximately satisfy the theoretical restrictions implied by consumer demand theory (Huang 1988).

Food producers see the own-price and quantity consumed of q and the data vector, G_t, which differs from X_t by a white noise observation vector, e_t, or:

$$G_t = X_t + e_t \tag{iv}$$

where $Ee_t = 0$, $Ee_t = \sigma^2_e$, and $Ee_t e_s = 0$ for all $s \neq t$. Hence, firms see:

$$P_t = -A_1 q_t + \beta' G_t + r_t \qquad (v)$$

where $Er_t = 0$, $Er_t = \sigma^2_r$, $Er_t r_s = 0$ for all $s \neq t$, and $Ee_t r_s = 0$ for all s and t. Substituting equation (iv) into (v) expresses the observed demand function as:

$$P_t = -A_1 q_t + \beta' X_t + \beta' e_t + r_t. \qquad (vi)$$

Defining $\theta_t \equiv \beta' X_t$ as the composite demand shifter, $u_t \equiv \beta' e_t + r_t$ as the composite observation error, it follows that $z_t = \theta_t + u_t$.

Appendix B. Starting Values

The GML estimates maximize a nonlinear optimization procedure, so the local optima depend on the initial values of ρ, σ_e, and σ_u. In addition, to implement the Kalman filter, initial values for the first two conditional moments of the unobserved demand shift, θ, must be specified. Furthermore, initial values of the λ and d parameters in the nonlinear regression must be specified. The procedure for computing these initial values is discussed in this section.

Starting values for the fixed parameters ρ, σ_e, and σ_u are obtained by replacing the unobserved shifter, θ, with data on the shifter, z in equation (7) in the text. In particular, the parameter estimate of ρ_0 in the linear regression:

$$z_t = \rho_0 z_{t-1} + e_t \qquad (i)$$

is used to initialize ρ in the GML routine. The residuals of this regression are collected, and starting values for σ_e, and σ_u are computed as:

$$\sigma_{\varepsilon,0} = \sigma_{u,0} = \left(\frac{1}{n} \sum_{t=1}^{n} e_t^2 \right)^{1/2}. \qquad (ii)$$

To implement the Kalman filter, it is required to initialize the conditional mean $E(\theta_0 | \Omega_{-1})$ and the conditional variance, $P_{0|-1}$ of the unobserved demand shift. Since the data are (fourth) differenced prior to estimation, $E(\theta_0 | \Omega_{-1})$ is set to the mean of the data, namely zero. As noted by Newbold and Bos (1985), prior to observing any data, the variance matrix cannot be updated. Hence the conditional variance is a constant. Denote $P_{0|1} = P_{-1|-2} = P$. The value of P is found by setting this constant equal to zero and iterating. Hence the initial value of P is found by first setting P equal to any value and iterating on:

$$K = P(P + \sigma_u^2)^{-1} \qquad (iii)$$

$$P = (\rho - \rho K)^2 P + \sigma_\varepsilon^2 + \rho^2 K^2 \sigma_u^2. \qquad (iv)$$

until the value of P converges. Note that the initial variance term, P and the initial Kalman gain term, K, are functions only of the parameter values ρ, σ_e, and σ_u. To start the GML optimization procedure, P and K are found by evaluating equations (iii) and (iv) at the initial values of ρ, σ_e, and σ_u. For subsequent Gauss-Newton steps, P and K must again be initialized. In this case the equations (iii) and (iv) are evaluated at the starting values of the Gauss-Newton step, namely the values of ρ, σ_e, and σ_u, computed in the previous step.

Finally, starting values for the λ and d parameters are computed. This is accomplished by computing OLS estimates of the parameters in the linear relationship:

$$f_t = b_1 f_{t-1} + b_2 z_t + \zeta_t.$$

b_1 is used as the starting value of λ. Setting $\beta = .95$ and $\alpha = 1$, the inital value of d, d_o, solves:

$$b_2 = \left(\frac{\alpha \lambda_o}{d_o [1 - \lambda_o \beta \rho_o]} \right)$$

14

Changes in Firm Behavior and Alternative Trade Policy Instruments

Steve McCorriston and Ian M. Sheldon

Introduction

In recent years, research focussing on modelling agricultural trade issues has been an important activity for many economists. The most notable attempts -- for example, those by OECD (1987), Roningen and Dixit (1990), and Tyers and Anderson (1992) -- have increased awareness of the costs of agricultural support policies and have identified the likely winners and losers from agricultural policy reform. However, one failing of these empirical models is that they ignore issues associated with market structure, assuming that agricultural produce is directly transferred from farmers to consumers, thus omitting the existence of the food processing and distribution industries. Further, given the structural and behavioral characteristics of these industries (see Sutton 1991), by making this exclusion, the role of market structure and its potential influence on the outcome of policy reform is ignored.

Recent computable general equilibrium (CGE) modelling efforts have improved this situation since they allow the capture of the intermediate nature of agricultural output. For example, Hertel and Lanclos (1992) have explicitly identified a role for the food processing sector in a multi-region CGE model, and, in so doing, enable a fuller specification of industry characteristics. There is a role for economies of scale, product differentiation, free entry and exit, and strategic interaction. In such an environment, agricultural policy reform can generate gains in excess of standard partial equilibrium models, not just due to the general equilibrium effects, but also due to the fact that trade reform can give rise to pro-competitive effects because of changes in market structure.

However, despite these recent improvements, one issue has not been considered: this is the impact different trade policy instruments have on the effects of policy reform. As is well known, governments use a variety of trade intervention measures to sustain high domestic prices, including import quotas, voluntary export restraints, variable import levies and tariffs and the like. In the context of partial equilibrium studies that ignore market structure issues, ignoring the type of instrument used is not a problem. But when dealing with oligopolistic markets, the issue is a relevant one since tariffs and quantitative restraints can have different effects, because the latter alters firms' behavior while the former leaves it unchanged. Consequently, when modelling imperfectly competitive markets, it may be insufficient to deal with trade policy instruments only in their *ad valorem* form.

Therefore, the effect of alternative trade policy instruments in oligopolistic markets is the focus of this chapter. It is shown that an import quota permitting the same level of imports that would enter under a corresponding tariff can have a very different outcome relative to the tariff policy. This issue is discussed with reference to the recent, and on-going, policy debate with respect to changes to the European Community's (EC) banana import regime, where the markets in member states are dominated by a small number of firms.

This chapter is organized as follows. We first review the literature on the tariff-quota (non-) equivalence issue and the role that market structure may play in influencing the outcome. A general conjectural variations model that captures the anticipated effects of tariffs and quotas in oligopolistic settings is then specified. The recent debate on the EC banana regime is discussed, and the effects of tariffs and quotas on banana imports by one EC member state (Germany) are evaluated. The final section summarizes and concludes.

Tariffs, Quotas, and Firms' Behavior

The literature on the equivalence between tariffs and quotas as alternative trade policy instruments has a long history. Originating with Bhagwati (1965), a series of papers (including, among others, Bhagwati 1968; Shibata 1968; and Sweeney, Tower, and Willett 1977) showed that market structure was important in determining whether tariffs and quotas were equivalent. 'Equivalence' relates to the effect on domestic prices when the level of imports permitted with a quota is the same as that arising with a tariff. Tariffs and quotas are equivalent if the effect on domestic prices is identical.

This early literature considered two extremes of market structure: perfect competition and monopoly. The upshot of this research was to show that market structure matters since, given that the quota-induced demand curve facing the monopolist lies below the initial demand curve, domestic prices will be higher with a quota relative to a tariff. That is, tariffs and quotas are non-equivalent under monopoly.

More recent literature has considered the tariff-quota issue in intermediate cases, i.e. where markets are oligopolistic. For example, Harris (1985), and Krishna (1989) have considered the effects of quantitative restraints when firms play Bertrand strategies. Krishna's (1989) paper is the more notable of the two, showing that, even if a quota is set at the free trade level, domestic prices rise since a quota facilitates anti-competitive behavior by domestic and foreign firms. In a quantity-setting framework, Digby, Smith, and Venables (1988) show that, with Cournot behavior, domestic prices will again be higher when quantitative restraints apply. This arises since the profit-maximizing condition for the domestic firm takes account of the constraint imposed on its competitors. Since Digby *et al.* (1988) model the quota constraint as a specified share of the market for foreign competitors, they argue that a 'pure' Cournot assumption would be irrational for home firms to make since extra sales by the home firm will increase sales of its foreign competitors. This gives rise to an anti-competitive effect since the level of sales of both firms will now be lower with a quota relative to the tariff, and domestic prices will be relatively higher.

Still within a quantity-setting framework, Hwang and Mai (1988) suggest a different framework. In a general conjectural variations model, quantitative restraints can generate either pro- or anti-competitive effects, or neither, depending on the initial values of the firms' conjectures--how (un)competitive the market was prior to the imposition of the quota. In their duopoly model, a quota effectively imposes Cournot behavior on the home firm. The home firm now knows that if it changes its output, the foreign firm's output cannot change, presuming the quota is binding, which is effectively identical to what is assumed under Cournot behavior. Hence, if the firm was initially playing more (less) competitively than Cournot, then a quota will make the market less (more) competitive. In this context, the quota can have either pro- or anti-competitive effects.[1] This outcome differs from that of Digby *et al.* (1988) where only anti-competitive effects can arise.[2]

In sum, when markets are oligopolistic, tariffs and quotas are likely to be non-equivalent since quotas affect firms' behavior while tariffs do not.[3] Consequently, when modelling imperfectly competitive markets, dealing with alternative trade policy instruments in *ad valorem* form ignores a potentially important effect of trade policy intervention. How important this issue is and what influences the degree of non-equivalence is explored in the remainder of this chapter.

Theoretical Framework

The model of oligopoly used here is a standard model of differentiated oligopoly utilized by, among others, Cheng (1987), Dixit (1988a), and Levinsohn (1989). There are two principal reasons for adopting this model: first, it follows a general conjectural variations approach so that the effect of quotas on firms' conjectures can be considered and second, following Dixit (1988b), the model can be used to generate an empirical assessment of the effect of trade policies.

The structure of the market is divided into two, where dominant firm(s) compete with fringe firms in the domestic market, although there is no domestic production. The dominant firm(s) output is denoted by subscript 1, fringe output by subscript 2. Consumer surplus is given by:

$$\Gamma = f(Q_1, Q_2) - p_1 Q_1 - p_2 Q_2 \tag{1}$$

where the utility function $f(Q_1, Q_2)$ is defined as:

$$f(Q_1, Q_2) = a_1 Q_1 + a_2 Q_2 - (b_1 Q_1^2 + b_2 Q_2^2 + 2k Q_1 Q_2)/2 \tag{2}$$

These functional forms generate the following inverse demand functions:

$$p_1 = a_1 - b_1 Q_1 - k Q_2 \tag{3}$$

$$p_2 = a_2 - b_2 Q_2 - k Q_1 \tag{4}$$

where all the parameters are positive, p_1 and p_2 are prices, Q_1 and Q_2 are quantities, and $b_1 b_2 - k^2 \geq 0$ indicates the extent to which dominant and fringe goods are imperfect substitutes.

On the supply side, there are n_i firms in the dominant and fringe sectors. Since in the example to follow the market is characterized by the dominance of multinational firms, it is assumed that profits are repatriated abroad. Thus, profits do not enter the domestic welfare function. Firms' costs are assumed constant. Profits for a representative firm in each sector are given by:

$$\pi_1 = (p_1 - c_1 - t)q_1 - f_1 \tag{5}$$

$$\pi_2 = (p_2 - c_2 - t)q_2 - f_2 \tag{6}$$

where c_i and f_i are marginal and fixed costs, respectively, and t is a tariff. Since there are n_i firms in each sector, such that aggregate output can be given by $Q_i (= n_i q_i)$, the first-order conditions for profit maximization are:

$$P_1 - c_1 - t - Q_1 V_1 = 0 \tag{7}$$

$$P_2 - c_2 - t - Q_2 V_2 = 0 \tag{8}$$

where the aggregate conjectural variation parameters V_i are explicitly given by:

$$V_1 = [D_1^1 \{1 + (n_1 - 1)v_{11}\} + D_2^1 n_2 v_{12}] / n_1 \tag{9}$$

$$V_2 = [D_1^2 \{1 + (n_2 - 1)v_{22}\} + D_2^2 n_1 v_{21}] / n_2 \tag{10}$$

where D_i^j (i = 1,2) are the partial derivatives $(\delta p_i / \delta Q_i)$ with respect to dominant and fringe firm produced goods, and v_{ii} (i = 1,2) are the firms' conjectures about how dominant and fringe competitors will respond to a change in quantities. The values for the v_{ii}'s are continuous variables whose values capture a range of possibilities concerning firm behavior. For example, if the dominant and fringe firms play Cournot strategies, then all v_{ii}'s will equal zero. Hence, the values of V_1 and V_2 will equal D_1^1/n_1 and D_2^2/n_2 respectively. For conduct more competitive (less competitive) than Cournot, $v_{ii} < 0$ ($v_{ii} > 0$). In the limit, $v_{ii} = -1$, the competitive outcome, or $v_{ii} = 1$, the collusive outcome. Clearly firms can hold different conjectures about their competitors in the dominant and fringe sectors.

Equilibrium prices and quantities are obtained by combining (3), (4), (7) and (8) to give:

$$\begin{bmatrix} Q_1 \\ Q_2 \end{bmatrix} = \frac{1}{\Delta'} \begin{bmatrix} b_2 + V_2 & -k \\ -k & b_1 + V_1 \end{bmatrix} \begin{bmatrix} a_1 - c_1 - t \\ a_2 - c_2 - t \end{bmatrix} \tag{11}$$

$$\begin{bmatrix} P_1 \\ P_2 \end{bmatrix} = \begin{bmatrix} a_1 \\ a_2 \end{bmatrix} - \frac{1}{\Delta'} \begin{bmatrix} \Delta + b_1 V_2 & kV_1 \\ kV_2 & \Delta + b_2 V_1 \end{bmatrix} \begin{bmatrix} a_1 - c_1 - t \\ a_2 - c_2 - t \end{bmatrix} \tag{12}$$

where $\Delta = (b_1 b_2 - k^2)$ and $\Delta' = (b_1 + V_1)(b_2 + V_2) - k^2$.

The effect of an import quota can be broken down into two parts: an *ad valorem* effect and a firm behavioral effect. The *ad valorem* effect acts like the tariff since, for the same quantity of imports, consumer prices will rise. Thus, it can be readily shown from the above framework that:

$$\frac{dp_1}{dt} + \frac{dp_2}{dt} > 0 \tag{13}$$

Thus, domestic prices rise when a tariff is imposed. However, the quota can potentially change firms' behavior which, in this model, changes firms' conjectures. Thus, if the quota affects all firms (which is the case in the particular example to follow), then, for a given tariff, it can be shown that:

$$\frac{dp_1}{dv_{11} + dv_{12}}\bigg|_t + \frac{dp_2}{dv_{21} + dv_{22}}\bigg|_t > 0 \tag{14}$$

If all firms were initially playing Cournot (i.e. $v_{11} = v_{12} = v_{21} = v_{22} = 0$) then tariffs and quotas would be equivalent since there would be no change in firms' conjectures induced by the quota. However, if the conjectures initially differed from zero (i.e. $v_{11}, v_{12}, v_{21}, v_{22} \gtrless 0$), then tariffs and quotas would be non-equivalent since the quota now has the additional effect of making the conjectures take the Cournot value. Of course, each of the conjectures could initially be positive or negative (e.g. $v_{11}, v_{12} > 0$ and $v_{21}, v_{22} < 0$ or any other combination), then the net effect will be the sum of the subsequent changes.[4]

In general, four factors will influence the degree of non-equivalence: (i) the initial value of each of the firm's conjectures; (ii) the number of firms in both sectors of the market; (iii) the market share of each group of firms--Q_1 relative to Q_2; and (iv) the degree of product differentiation between the two products.

In terms of domestic welfare, therefore, quotas can exacerbate or offset the effects of tariffs. However, there are further considerations. With tariffs, there can be 'rent-shifting' effects. In most models, this relies on sufficiently strong terms of trade effects, though in the example to follow, it could arise if the government collects tariff revenue from the dominant and fringe firms (since neither of these are domestic producers). Thus, defining welfare to be the sum of consumer surplus and tariff revenue on all imports (Q_1 and Q_2), then it is possible that:

$$\frac{dW}{dt} > 0 \tag{15}$$

However, with quantitative restraints, unless the quota licenses are auctioned, the quota rent would not be retained domestically. In this case, domestic welfare (now defined only as consumer surplus) falls.

In sum, quotas in oligopolistic markets have three potential effects: (i) they have an *ad valorem* effect similar to that of a tariff; (ii) they create quota rents which may not be retained domestically; and (iii) they may change firms' behavior and can, therefore, have pro- or anti-competitive effects.

The EC Banana Market

Throughout the EC, the world's largest market for bananas, individual member states apply a range of restrictions regarding the imports of bananas. With no domestic production of its own (except Greek imports from Crete), the purpose of these import restrictions is to maintain access to certain EC member state markets and ensure high prices for preferential suppliers. These preferential suppliers are the banana exporting countries of the African, Caribbean and Pacific (ACP) states whose preferential treatment is embodied in the EC-ACP Lomé Convention. Under Protocol 5 on Bananas of the Lomé Convention, Article 1 states that:

> In respect of its banana exports to Community Markets, no ACP state shall be placed, as regards access to its traditional markets and its advantages on those markets, in a less favorable position than in the past or at present.

Other countries that retain preferential access to the EC are the Overseas Departments of France (e.g. Guadeloupe), Spain (Canary Islands) and Portugal (Madeira).

The means by which preferential access and high prices are sustained in certain member countries is done by restricting the imports of bananas from so-called Dollar countries (e.g. Columbia and Ecuador). So, for example, in the UK, an import quota and a tariff of 20 percent apply on imports from Dollar countries, while ACP bananas, which typically account for 75 percent of UK banana supplies, enter unrestricted. In France, Greece, Portugal and Spain, imports of Dollar bananas are restricted and an import duty applies. In Belgium, Denmark, Luxembourg, Ireland, and the Netherlands, who have no tradition of banana exporting colonies, a 20 percent tariff is applied, but most imports come from Dollar countries. Finally, Germany, the largest single banana market in the EC, has no restrictions on banana imports, with most supplies coming from Dollar countries.

As a result of differing restrictions on banana imports between EC member states, the retail prices of bananas vary substantially throughout the EC. Table 14.1 gives an indication of the extent of this price variation. Germany has the lowest prices in the EC, which is unsurprising given the absence of trade barriers. Italy has the highest prices, more than double those in Germany, while in France, the UK and Spain retail prices range between 60-70 percent above those in Germany. For the rest of the EC, prices are 20 percent above German prices.

Such price differences between member states are sustained by Article 115 of the Treaty of Rome which, under prescribed circumstances, can be used to restrict trade between member states when one (or all) EC member state(s) apply national restrictions on imports from third countries. By restricting such trade flows, this prevents price harmonization in the EC banana market.

This situation is, however, unsustainable. With the 1992 Single Market Process, preventing such trade flows is inconsistent with the spirit of the '1992' program, and impossible to maintain given the absence of customs posts. Thus, the problem facing the EC Commission has been to formulate a new banana import regime, and one that will be common to all EC member states. However, given its obligations under the Lomé Convention, the Commission's options are constrained since it is legally bound to ensure that those countries with preferential access are not made worse off by any changes to the EC's banana regime (see Protocol 5 above).

In the course of this debate, a series of studies have appeared analyzing the options open to the EC.[5] While the optimal policy would likely have been a free trade regime with direct compensation to current preferential suppliers, this option has not been given serious consideration by the EC Commission. Rather, the main options revolved around setting either EC-wide tariffs or quotas, or some combination of

TABLE 14.1 EC Retail Prices for Bananas (1987) (Germany = 100)

Germany	100
Italy	211
France	173
UK	168
Spain/Portugal	157
Other EC*	120

Note: * Belgium, Denmark, Luxembourg, Ireland, and Netherlands.
Source: Fitzpatrick and Associates (1990).

the two. While there is obviously some variation between the empirical studies, most are fairly consistent with each other: the appropriate level of the tariff that would be consistent with the EC's obligations would be in the range 17-25 percent, with the commensurate EC-wide quota being in the range of 2 million metric tons. With such changes, German consumers would be the biggest losers, with some consumer gains in the UK, France, Spain and Italy. Preferential suppliers would be unharmed by such changes.[6]

However, one particular problem with these empirical assessments is that they ignore relevant characteristics of the structure of the EC banana market. In all the empirical models, perfectly competitive market structures are implicitly assumed. This assumption, however, does not fit with the facts. The banana market worldwide is dominated by a few multinationals, so that the wholesaling and distribution of bananas is highly concentrated.[7] Table 14.2 shows that three multinationals -- United Brands (who are responsible for Chiquita), Dole (who control Standard Fruit) and Del Monte -- account for 70 percent of the world market and 66 percent of the European market.

In individual EC states, markets are also highly concentrated. Table 14.3 reports the market shares for the top three banana suppliers in Italy,France and Germany. In Italy, the three main suppliers (Chiquita, Comafrica, and Simba) account for 68 percent of the market; in France, the three-firm concentration ratio is 51 percent (involving Pomona/Benexo, Compagnie Frutiere, and CDB Dunand); in Germany, three firms (Chiquita, Dole, and Noboa) account for 72 percent of the market.

Given such characteristics, the market structure of the banana sector in EC member states is more accurately characterized as being oligopolistic. Consequently, in light of the discussion earlier when considering the impact of an import quota, it is inaccurate to just find its

TABLE 14.2 Multinationals in the World Banana Market (1990)

	Share of World Market (%)	Share of European Market (%)
United/Chiquita	35	43
Standard/Dole	20	13
Del Monte	15	10

Source: Hallam and McCorriston (1992).

TABLE 14.3 Market Shares for the Three Main Banana Suppliers in Italy, France and Germany

Italy (1988)	%	France (1989)	%	Germany (1989)	%
Chiquita Italia	36	Pomona/Benexo	28	Chiquita	40
Comafrica	18	Compagnie Frutiere	12	Dole	16
Simba	14	CDB/Dunand	11	Noboa	16

Source: Fitzpatrick and Associates (1990).

tariff-equivalent and estimate the corresponding welfare effects. Such trade instruments may also change firms' behavior. An assessment of these effects is made in the following section.

Empirical Assessment

Methodology

The methodology for evaluating the effects of tariffs and quotas in oligopolistic markets requires calibrating the demand system in equations (3) and (4) for a particular market. Calibrating this system involves using estimates of the elasticity of demand, the elasticity of substitution and observations on prices, quantities and costs so that the parameters are consistent with equilibrium in any given period.[8] Once this is done, the effects of policy changes can be derived using equations (11) and (12), the changes in consumer surplus being found using equation (1).

Data

This methodology is applied to the German banana market, the EC member state that has operated a free market system. The data necessary to calibrate equations (3) and (4) are given in Table 14.4.

The dominant firm (subscript 1) is taken as Chiquita with the remaining suppliers taken as the fringe (subscript 2). Price data have been taken from FAO Banana Statistics (1992), as have estimates of costs, taken to be the landed price. Quantity data comes from Borrell and Yang (1992). The elasticity of demand is - 0.4 as reported in Islam and Subramian (1989). No data are available for the elasticity of substitution. While bananas from alternative sources can be regarded as good

TABLE 14.4. Calibration Data for the German Banana Market (1989)

Q_1	321,925 metric tons
Q_2	493,075 metric tons
P_1	1378 DM/metric ton
P_2	1183 DM/metric ton
c_1	1002 DM/metric ton
c_2	1002 DM/metric ton
ε	- 0.4
σ	3.0

Source: Islam and Subramian (1989), Borrell and Yang (1992), FAO Banana Statistics (1992).

substitutes for each other, there are perceived quality differences. Hence, a value for the elasticity of substitution was initially taken to be 3.0. However, given that product differentiation is a potentially important influence on the non-equivalence between tariffs and quotas, a sensitivity analysis of the results to this value is carried out.

Results

The EC Commission's policy options varied between tariffs and quotas. It is assumed that the EC-wide tariff chosen is 20 percent or, alternatively, that the EC imposes quantitative restraints on banana imports at the 20 percent tariff-equivalent level. The effects of these policies on the German banana market are given in Table 14.5. As expected, the imposition of the tariff would reduce German consumers' welfare, the estimated reduction here being DM 146 million, a reduction of approximately 11 percent. Overall welfare for Germany is reduced by this amount even though the 20 percent tariff raises revenue. This, however, becomes part of EC budgetary resources. On the other hand, if it were retained in Germany, welfare shows a small net increase. Firms supplying the German market would face a reduction in profits of DM 22 million.

Importantly though, as discussed above, the quota-equivalent policy is likely to have different effects since not only will it have an *ad valorem* aspect, but it will also change firms' behavior unless they are already playing Cournot. However, the calibration results suggest that this was not so.

Table 14.6 suggests that the dominant firm was playing more competitively than Cournot while the fringe firms (in aggregate) were

TABLE 14.5. Effect of Tariffs and Quotas on the German Banana Market (DM million)

Simulation	Consumer Welfare	Firms' Profits	Tariff Revenue
Baseline	1283.6	210.3	-
20% Tariff	1136.7	187.9	153.2
20% Tariff -Equivalent Quota	966.0	411.1	-

TABLE 14.6. Estimated Conjectural Variations[*]

	Actual	Cournot-Equivalent
Dominant Firm (V_1)	.0016800	.0086751
Fringe Firms (V_2)	.00036708	.00030505

Note:[*] Calculated from calibration using equations (9) and (10).

playing slightly less competitively than Cournot. Consequently, with the imposition of a quota, the dominant firm now plays less competitively, the fringe more competitively. The net effect, however, is that the market becomes less competitive. Domestic prices, therefore, rise by a greater amount than in the tariff case and consumers' welfare is reduced further. Table 14.5 shows that, relative to the baseline, consumer welfare is reduced by DM 317.6 million, a reduction of 25 percent. This is more than double the losses resulting from the 20 percent tariff case. Further, unless the quota licenses are auctioned, quota rents are captured by the supplying firms whose profits now increase by 95 percent relative to the baseline case and by 118 percent relative to the tariff case. Clearly, at least in this example, the non-equivalence between tariffs and quotas is substantial.[9]

Sensitivity Analysis

The degree of non-equivalence between tariffs and quotas in oligopolistic markets is likely to be sensitive to the degree of product differentiation. This section, therefore, explores this issue by varying the value for the elasticity of substitution used in the calibration. To do this, bounds are fixed on the distribution of σ, these bounds being set at 0.5 and 5.5, and the distribution of σ is assumed to be uniform. A random

FIGURE 14.1. Sensitivity Analysis of Welfare on σ

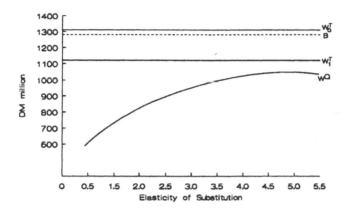

Key: B = Baseline Welfare, W_0^T = Welfare with Retention of Tariff Revenue, W_1^T = Welfare with no Tariff Revenue, W^Q = Welfare with Quota.

number generator is used to make 1000 draws from this distribution, and can retain the tariff revenue. In the other two cases, as before, the tariff revenue is transferred to the EC budget or, as in the case of the quota, the firms capture the quota rent.

Figure 14.1 plots the welfare changes for these three scenarios and also for the baseline case. If the German government retained the tariff revenue, German welfare would increase slightly as this revenue is collected from both suppliers. This 'rent-shifting' effect outweighs losses to consumers.   If the tariff revenue were transferred to the EC Commission, German welfare would be lower than the baseline level, the reduction in welfare being almost invariant to the extent of product differentiation. With the quota-equivalent, however, welfare would be at its lowest level, the difference between the tariff case, with no tariff revenue to the government, and the quota outcome being due to the anti-competitive effects. As Figure 14.1 shows, the degree of non-equivalence varies (inversely) with the elasticity substitution.   For relatively low values, the degree of non-equivalence rises and the anti-competitive effects are stronger than the *ad valorem* effects. Even if the German government retained the quota rent, welfare would still be substantially reduced.

Summary

This chapter has focused on the role of market structure in assessing the effects of trade policy. While there have been some recent attempts to deal with market structure issues in agricultural trade policy analysis, there has, as yet, been little discussion of the effects of alternative trade policy instruments. It has been shown that quantitative restraints in oligopolistic markets are likely to have different effects relative to the equivalent level of imports induced by a tariff. The reason for this is that a quantitative measure will change firms' behavior while a tariff will leave it unchanged.

This issue was explored in the context of the changes proposed to the EC's banana market regime. In the empirical assessment of trade measures applied to the German market, it was shown that the anti-competitive effects would outweigh the *ad valorem* and tariff revenue (if applicable) effects. However, it may be possible that pro-competitive effects could be the dominant feature in other case-studies. In sum, in analyzing the effect of trade policy in oligopolistic environments, it is necessary to take account of the trade instrument used. Focusing just on *ad valorem* equivalents does not capture the whole story. Indeed, it might not even capture the most important part of it.

Notes

1. If the firms were initially playing Cournot, the quota would not have any effects on firms' behavior.

2. Mai and Hwang (1989) show that prices are higher under a ratio quota than prices under a volume quota.

3. Other papers that deal with tariff-quota non-equivalence include: Itoh and Ono (1982, 1984), Fung (1989), and Levinsohn (1989).

4. Of course, it is possible that even with initial conjectures being different from one, the pro- and anti-competitive effects could cancel each other.

5. These include Borrell and Yang (1990, 1992), Fitzpatrick and Associates (1990), Borrell and Cuthbertson (1991), Hallam and McCorriston (1992), and Matthews (1992).

6. The policy that was actually introduced in December 1992 was a combination of tariffs and quotas. For the first 2 million metric tons, bananas will enter the EC at a reduced duty. Thereafter, the tariff rises to a prohibitive 170 percent.

7. Some multinationals are also involved in the production process.

8. Further details on the calibration procedure can be found in Dixit (1988b).

9. In their study of protection in the EC car market, Digby *et al.* (1988) also found that voluntary export restraints gave rise to considerable anti-competitive effects.

References

Bhagwati, J.N. 1965. "On the Equivalence of Tariffs and Quotas," in R.E. Baldwin ed., *Trade, Growth and the Balance of Payments-Essays in Honor of G. Haberler.* Chicago, Ill: Rand McNally.

_____. 1968. "More on the Equivalence of Tariffs and Quotas." *American Economic Review* 58:142-146.

Borrell, B. and S. Cuthbertson. 1991. *EC Banana Policy 1992.* Canberra: Centre for International Economics.

Borrell, B. and M.C. Yang. 1990. *EC Bananarama 1992.* Working Paper 523. Washington D.C.: World Bank.

_____. 1992. *EC Bananarama 1992: The Sequel.* Working Paper 958. Washington D.C.: World Bank.

Cheng, L. 1988. "Assisting Domestic Industries under International Oligopoly: The Relevance of the Nature of Competition to Optimal Policies." *American Economic Review* 78:743-758.

Digby, C., A. Smith, and A. Venables. 1988. *Counting the Cost of Voluntary Export Restraints in the European Car Market.* Centre for Economic Policy Research Discussion Paper No. 249.

Dixit, A.K. 1988a. "Anti-Dumping and Countervailing Duties under Oligopoly." *European Economic Review* 32:55-68.

_____. 1988b. "Optimal Trade and Industrial Policy for the U.S. Automobile Industry," in R.C. Feenstra, ed., *Empirical Methods for International Trade,* Cambridge, MA: MIT Press.

FAO. 1992. *Banana Statistics.* Rome.

Fitzpatrick, J. and Associates. 1990. *Trade Policy and the EC Banana Market: An Economic Analysis.* Dublin.

Fung, K.C. 1989. "Tariffs, Quotas, and International Oligopoly." *Oxford Economic Papers* 41:749-757.

Hallam, D. and S. McCorriston. 1992. "International Trade Policies in Bananas and Proposals to Alter Existing Policies in Line with the Single European Market." in J.P. McInerney and M. Peston, eds., *Fair Trade in Bananas?* Occasional Paper No. 239. University of Exeter.

Harris, R. 1985. "Why Voluntary Export Restraints are 'Voluntary'." *Canadian Journal of Economics* 18:799-801.

Hertel, T.W. and D.K. Lanclos, 1994. "Trade Policy Reform in the Presence of Product Differentiation and Imperfect Competition: Implications for Food Processing Activity," in M. Hartmann, P. Schmitt, and H. Von Witzke, eds., *Agricultural Trade and Economic Integration in Europe and North America.* Kiel: Wissenschaftsverlag Vauk Kiel KG.

Hwang, H. and C.C. Mai. 1988. "On the Equivalence of Tariffs and Quotas under Duopoly." *Journal of International Economics* 24:373-380.

Islam, N. and A. Subramian. 1989. "Agricultural Exports of Developing Countries: Estimates of Income and Price Elasticities of Demand and Supply." *Journal of Agricultural Economics* 40:221-231.

Itoh, M. and Ono, Y. 1982. "Tariffs, Quotas and Market Structure." *Quarterly Journal of Economics* 97:295-305.

_____. 1984. "Tariffs vs. Quotas under Duopoly with Heterogeneous Goods." *Journal of International Economics* 17:359-373.

Krishna, K. 1989. "Trade Restrictions as Facilitating Practices." *Journal of International Economics* 26:251-270.

Levinsohn, J. 1989. "Strategic Trade Policy When Firms Can Invest Abroad: When Are Tariffs and Quotas Equivalent?" *Journal of International Economics* 27:129-146.

Mai, C.C. and H. Hwang. 1989. "Tariffs versus Ratio Quotas under Duopoly." *Journal of International Economics* 27:177-183.

Matthews, A. 1992. *The European Community's Banana Policy after 1992.* Discussion Paper 13. University of Giessen: Institut für Agrarpolitik und Marktforschung.

OECD. 1987. *National Policies and Agricultural Trade.* Paris: OECD

Roningen, V.O. and P.M. Dixit. 1990. "Assessing the Implications of Freer Agricultural Trade." *Food Policy* 1:67-76.

Shibata, H. 1968. "A Note on the Equivalence of Tariffs and Quotas." *American Economic Review* 58:137-142.

Sutton, J. 1991. *Sunk Costs and Market Structure.* Cambridge, MA: MIT Press.

Sweeney, R.J., E. Tower, and T.D. Willett. 1977. "The Ranking of Alternative Tariff and Quota Policies in the Presence of Domestic Oligopoly." *Journal of International Economics* 7:246-262.

Tyers, R. and K. Anderson. 1992. *Disarray in World Food Markets – A Quantitative Assessment.* Cambridge: Cambridge University Press.

15

Optimal Auction Theory and EC Grain Exports

Jean-Marc Bourgeon and Yves Le Roux

Introduction

Within the Common Agricultural Policy (CAP), export refunds and import levies are the major tools of trade regulation with third countries. Two-thirds of European grain policy expenditures are concerned with these export refunds, which account for ten percent of the total agricultural budget.

For several years, major exporting countries have been fighting a world-wide war over cereals trade. Increasing grain exports are associated with increasing costs. The 1992 CAP reform aims to curb this evolution by bringing down internal prices to a level close to the world price. But such measures will have only indirect effects on trade. In particular, quantitative effects on trade cannot be anticipated precisely.

Export refund awarding procedures remain the only instrument that allows direct control of European exports and public costs. To bridge the gap between the domestic price and the world price, the Community can resort to several kinds of subsidies according to the origin of cereals (intervention stocks which are the property of the Community or private stocks) and their destination. It is also necessary to distinguish between standing refunds, which are generally available without limitation on quantity, and export refunds which are tendered for. In the latter case, competition between traders simultaneously settles the unitary refund level and the quantity to be exported. The Community has used these different procedures in various proportions over time, thus indicating its strong intention to apply the most efficient -- the least costly -- export policy.

This chapter analyzes European export refund awarding. Our analysis is focused on export refunds which are tendered for. The first section provides a detailed description of both this awarding policy and the behavior and objectives of the different agents (the European Commission and traders). Refund tenders are auction mechanisms, so the next section summarizes auction theory which can be used to model these awarding instruments. The third section is about the econometrics of auctions, and presents different methods associated with appropriate paradigms. Lastly, we present an empirical application of these methods to tenders involving soft wheat intervention stocks.

Awarding of Cereal Export Refunds

In the internal market, the European Community (EC) sets a market floor (or intervention) price, which is in fact a target price for intra-EC trade. Farmers can sell their products to the intervention authorities at this annually adjusted intervention price. Grain held in intervention stocks is disposed of on the domestic market or through exports.

The world price is essentially the outcome of the rivalry among major exporting countries through their export subsidy policies. Generally, the world price is less than the Community price. To ensure its export competitiveness, traders are given a refund, which makes up for the difference between the world price and the internal EC price. This refund is not a simple and direct compensation, because the world price is the result of subsidies awarded by the EC and other countries (the Export Enhancement Program (EEP) in the United States, for example).

The political stakes of refund awarding by the EC are obvious. On the one hand, it is critical at the international level, because winning or saving market shares can be achieved by higher refunds. On the other hand, the cost of such a policy is a burden on the Community budget. Therefore, refund awarding is of great importance in the management of the Community's export policy.

Instruments used by the European Commission are of two classes. First, there is a fixed or standing refund, for which the unitary subsidy is constant and can be awarded for any quantity. Secondly, the EC sets refunds and quantities according to traders' bids on tenders. (ONIC 1991; CAP Monitor 1993).

Initially, the CAP only provided for the use of standing refunds. However, their fixity involved speculative behavior when the world price fluctuated too much. In such cases, it was not possible to adjust the refund immediately. Moreover, it was difficult to control exported quantities within this standing refund procedure. So this procedure was gradually given up in favor of tenders. Tenders facilitate competition among traders and consequently reduce the unitary refund. In addition, tenders simultaneously allow control of the export price, which reflects the commercial policy of the Community, and of quantity (price and quantity finally determine the budgetary cost).

Tender Procedures

There are two tender procedures which, in the case of soft wheat, are used for 50 to 70 percent of European exports. The sharing of tenders between intervention stocks and the open market depends on both the world and European markets. Thus, high levels of intervention stocks and a weak absorption capacity in the internal market lead the EC to dispose of its stocks on the world market. Quantities offered by open market tenders are fixed as a residual.

Tenders for Intervention Stocks: Intervention stocks that the EC wants to export are tendered by lots. Tenders take place each week and concern several lots at a time. For one lot, each trader makes a per (metric) ton sealed-bid. The bidding price is determined keeping in mind the necessity of selling the lot on the world market at a competitive price.

In practice, the European Commission retains the highest bid for each lot. Because several lots are involved, the highest bids are grouped together and ranked. The Commission fixes a floor level, and all bids higher than this level are accepted. This level corresponds to the minimum price the Commission judges acceptable, based on offers made. At the same time, it corresponds to the maximum quantity the EC wants to export.

For simplicity, we will consider that a separate tender applies to each lot. For a given lot the highest bid is accepted if this bid is higher than a minimum sale price, or reservation price. This reservation price agrees with Commission objectives in terms of total quantity to be exported and in terms of minimum export price.

Open Market Tenders: In the case of open market tenders, the Commission buys an export service from traders who must stock up on grain from within the internal market. Bids are anonymously submitted to the Cereals Management Committee (which executes the Commission policy in the matter of cereals). The tenderer must specify the quantity he intends to export and the desired export refund. This refund reflects the subsidy level which is necessary to export to the world market, according to each tenderer (that is, as a function of his supplying cost within the internal market).

Each week the Committee lists the traders' proposals in increasing order of desired refunds, and adds up the associated quantities. Then, the Committee fixes a maximum export refund according to the export price that refund involves and the corresponding total quantity to be exported. A contract is awarded to any tenderer who has tendered a rate of refund equal to or less than the maximum refund. Traders get the proposed refund (not the maximum one) if the obligation to export is fulfilled.

Behavior of Traders and of the Commission

Whatever procedure is used by the Commission, three criteria are taken into account: the quantity to be exported (for a week, and for the whole marketing year), the budgetary cost of export financing, and the Community export policy (that is, its willingness to save or gain market share). That leads the Commission to anticipate various key variables: world supply and demand evolution, subsidies of other exporting countries, world prices, and Community public and private stocks. At the same time, traders have to form expectations on the same variables. According to these expectations, they determine their optimal bidding behavior, in terms of refunds for the open market or in terms of purchase price for intervention stocks. Trader's behavior can also be a function of information stemming from previous tenders, and can entail a speculative component. Getting a refund is closely connected to purchasing and selling of export licences, and consequently of export contracts.

Within a given tender procedure, the objective of this research is to determine and to model optimal behavior of traders according to previous mechanisms. It is also to highlight the nature of the traders' information -- that is, their valuation of the tendered object. Two opposite assumptions can be considered. First, the object has a different and certain value for each bidder. Each bidder considers his own value when bidding. This assumption is referred to as the private value paradigm. Generally one assumes that these individual values are independent of each other. Conversely, the second assumption is that the object has a common value for all bidders *ex post*, but this value is unknown *ex ante*. Before bidding, traders simply form an estimate of this common value. Among traders, estimates are different according to the information they have.

A bidder determines his bidding strategy according to his own estimate and to his prior belief about the others' estimates. The nature of the information determines the framework of behavior modelling. Knowledge of which paradigm applies is particularly important for public authorities. This knowledge would provide them with the information necessary to apply suitable tender awarding procedures.

Drawing a conclusion about the validity of one of the two competing assumptions gives information about traders' valuation of grain in the world market. At least it gives information about the valuation they take into account when bidding. If the private value assumption turns out to be correct, that means possible market segmentation exists for the same object. Hence, each trader can have specific market power (according to the specific geographical area he currently serves, for example). In the case of the world grain market, trade can be competitive, but there are bilateral agreements too, where political considerations prevail over market clearing. Existence of private values can also be the result of speculative behavior. On the one hand, traders compete to get export contracts. On the other hand, they contest for export refunds to hold these contracts. An imbalance between these two markets can produce different valuations for the same object according to traders.

The coexistence of these two markets does not exclude the common value assumption. In the case of common valuation, that would mean an equilibrium is established between these two markets, on average. The common value assumption involves a single valuation by traders when bidding. But it is not necessarily proof of the "law of the one-price" on the world market. It just means that traders determine their optimal bidding as if the object has – *ex-post* – the same value.

Auction Modelling

This section is devoted to a brief presentation of auction theory within the two competing paradigms of private and common valuation of the bidded object. For a more detailed description, an interested reader can refer to the McAfee and McMillan (1987) survey. In the following, we refer to the "principal" as the designer of the auction who wants to sell a single object and "bidders" as the possible opponent buyers. (We assume that there is no coalition among buyers. For an analysis of this point, see Graham, Marshall and Richard 1990). An auction is compared to a game, defined by an allocation rule (in most cases, the principal sells the object to the bidder who has made the highest proposal) and a payment rule. (For example, he can decide that the price will be the highest announced bid. In this case, he organizes a first price auction; or the price can be the second highest bid). The problem is to determine the expected behavior of a risk neutral bidder i confronted with these rules. We shall consider, for simplicity, a first price sealed bid auction.

The literature on auctions has focused on two opposite approaches concerning a bidder's individual valuation of the object. We shall derive optimal strategies within these two paradigms, with the extra assumption that the bidders are identical (symmetry assumption).

Private Value Paradigm

Within the private value assumption, an individual i, if he wins the auction (i.e. his bid is the highest one), will gain from the difference between his announced bid, b_i, and his valuation of the object, v_i. Consequently, his gain Π_i depends on his private valuation v_i, his bid b_i, and the total set of the announced bids $b_{-i} = (b_j)_{j\neq i}$. The expression for this profit for particular values of v_i, b_i, b_{-i} is:

$$\Pi_i(v_i, b_i, b_{-i}) = [v_i - b_i]\, 1_{(b_i > b_j\, \forall j\neq i)} \tag{1}$$

where $1_{(\omega)}$ is the indicator function that equals 1 if ω occurs, 0 otherwise.

Each individual determines his bid b_i to maximize his surplus. While each bidder knows his own valuation of the item, this valuation is unknown by the other bidders, who, given the assumption of independently distributed private values, cannot rely on their own valuations to estimate it. For the other bidders, the private value of individual j is a random variable V_j. (We assume that the cumulative distribution function F_j of this random variable is common knowledge.) For a particular realization $(v_j)_{(j\,=\,1,...,n)}$ of the random variables $(V_j)_{(j\,=\,1,...,n)}$, a Nash equilibrium of the auction game is defined by n bids $(b_j)_{(j\,=\,1,...,n)}$, so that it is not in the interest of any bidder to individually modify his offer. Considering all possible values of the stochastic variable $(V_j)_{(j\,=\,1,...,n)}$, an auction game equilibrium is consequently defined by n bid functions $(b_j^*(.))_{(j\,=\,1,...,n)}$. To determine such an equilibrium, we shall characterize the bid functions $b_j(.)$. Intuitively, they are increasing functions of the individual valuations v_j:

$$b_j \equiv b_j(v_j) \text{ and } b_j'(.) > 0\ (\forall j = 1,...,n)$$

Given that private values are stochastic, individual bids are random variables denoted by $B_j = b_j(V_j)$. Furthermore, if the bidder i is risk neutral, he will be interested in his expected surplus:

$$
\begin{aligned}
E\Pi_i(V_i, b_i, B_{-i}) &= E[[v_i - b_i]\, 1_{(b_i > b_j\ \forall j\neq i)}] \\
&= [v_i - b_i]\, E[1_{(b_i > b_j\ \forall j\neq i)}] \\
&= [v_i - b_i]\, P[\cap_{j\neq i}(b_i > B_j)]
\end{aligned}
$$

$P[\omega]$ denotes the probability of the event ω. If the other bidders follow their equilibrium strategies, B_{-j}^*, this expected surplus becomes:

$$E\Pi_i(v_i, b_i, B_{-i}^*) = [v_i - b_i] P[\cap_{j \neq i}(b_i > B_j^*)]$$

Under the symmetry assumption, the distribution of private values of equal bidders are the same, and they adopt the same equilibrium strategies:

$$\begin{cases} F_j(.) = F(.) \ (\forall j = 1,...,n) \\ B_j^* = b^*(V_j) \ (\forall j = 1,...,n) \end{cases}$$

Increasing b^* has an inverse function b^{*-1}. When applied to stochastic equilibrium bids, B_j^*, it generates the private value stochastic variable V_j. Consequently:

$$E\Pi_i(v_i, b_i, V_{-i}) = [v_i - b_i] P[\cap_{j \neq i}(V_i < b^{*-1}(b_i))] \tag{2}$$

where, in the particular case of independently and identically distributed stochastic variables:

$$P[\cap_{j \neq i}(V_i < b^{*-1}(b_i))] = \Pi_{j \neq i} P(V_j < b^{*-1}(b_i)) = [F(b^{*-1}(b_i))]^{n-1}$$

Expected profit of the individual i, when the other bidders follow their equilibrium strategies, is:

$$E\Pi_i(v_i, b_i, V_{-i}) = [v_i - b_i] [F(b^{*-1}(b_i))]^{n-1}$$

Assuming that $b^*(.)$ is differentiable, a necessary condition for a bid $b_i = b^*(v_i)$ to be optimal is provided by:

$$\left. \frac{\partial E\Pi_i(v_i, b_i, V_{-i})}{\partial b_i} \right|_{b_i = b^*(v_i)} = 0$$

That is:

$$[v_i - b^*(v_i)] \frac{d}{dv} [F(v_i)]^{n-1} - b^{*\prime}(v_i)[F(v_i)]^{n-1} = 0,$$

which is a first order differential equation, the solution of which is:

$$b^*(v_i) = v_i - \frac{\int_{v_0}^{v_i} [F(v)]^{n-1} dv}{[F(v_i)]^{n-1}} \tag{3}$$

where v_0 is the principal reservation value (the minimum value of acceptable bids).

Common Value Paradigm

Now consider the case where the principal proposes an item which is worth the same value c for all bidders (for example, its resale market value). This value is unknown at the time of the auction. Each bidder is supposed to be able to calculate a personal estimation of the common value c. We denote a_i as the realization of the random variable A_i of such an estimate by the individual i, and assume that all these expectations are conditional upon the *ex post* common value c.

In competing in a sealed bid auction, a bidder i cannot observe the behavior of others and therefore cannot improve his own valuation during the process of the game. However, if he presumes that his offer b_j is the highest one, he is obliged to attach probabilities to other bidders' estimated values. That induces refinements of his own valuation of the object (which is not the case when valuations are independently and identically distributed).

The surplus of bidder i for particular values (c, a_{-i}, b_{-i}) of the associated random variables is:

$$\Pi_i(c, a_i, a_{-i}, b_i, b_{-i}) = [c - b_i] 1_{\{b_i > b_j \ \forall j \neq i\}}$$

Without any information about the item and the others' valuations, he has to realize an estimate of the item. Hence, his surplus turns to (equilibrium is not assumed):

$$\Pi_i(C, a_i, A_{-i}, b_i, b_{-i}) = E[C - b_i | A_i = a_i, b_{-i}] 1_{\{b_i > b_j \ \forall j \neq i\}}$$

His expected profit is:

$$E\Pi_i(C, a_i, A_{-i}, b_i, B_{-i})$$
$$= E[E[C - b_i | A_i = a_i, B_j < b_i \ \forall j \neq i] 1_{\{b_i > B_j \ \forall j \neq i\}} | A_i = a_i]$$

Assuming that other bidders apply their equilibrium $B_{-i}^* = b^*(A_{-i})$, the expected profit becomes:

$$E\Pi_i(C, a_i, A_{-i}, b_i, B_{-i}^*)$$
$$= E[E[C - b_i \mid A_i = a_i, B_j^* < b_i \; \forall j \neq i] \, 1_{\{b_i > B_j^* \; \forall j \neq i\}} \mid A_i = a_i]$$

Assuming symmetry and applying the inverse function b^{*-1} to the equilibrium bids, this becomes:

$$E\Pi_i(C, a_i, A_{-i}, b_i, B_{-i}^*)$$
$$= E[E[C - b_i \mid A_i = a_i, A_j < b^{*-1}(b_i) \; \forall j \neq i] \, 1_{\{A_j < b^{*-1}(b_i) \; \forall j \neq i\}} \mid A_i = a_i]$$

Inverting the integration order:

$$E\Pi_i(C, a_i, A_{-i}, b_i)$$
$$= E[E[1_{\{A_j < b^{*-1}(b_i) \; \forall j \neq i\}} \mid A_i = a_i](C - b_i) \mid A_i = a_i, A_j < b^{*-1}(b_i) \; \forall j \neq i]$$

This expression can be reformulated in the case of i.i.d. stochastic variables:

$$E\Pi_i(C, a_i, A_{-i}, b_i)$$
$$= \int [F(b^{*-1}(b_i) \mid C = c)]^{n-1} (c - b_i) \, dP[C \mid A_i = a_i] \tag{4}$$

Assuming $b^*(.)$ is differentiable, a necessary condition for the bid function $b^*(.)$ to be a symmetric equilibrium strategy is:

$$\left. \frac{\partial E\Pi_i(C, a_i, A_{-i}, b_i)}{\partial b_i} \right|_{b_i = b^*(a_i)} = 0$$

Which gives:

$$b^{*\prime}(a) = \frac{\int [c - b^*(a_i)](n-1) f(a_i \mid C = c)[F(a_i \mid C = c)]^{n-2} dP[C \mid A_i = a_i]}{\int [F(a_i \mid C = c)]^{n-1} dP[C \mid A_i = a_i]} \tag{5}$$

In the following, our interest is in applying Thiel's (1988) model, discussed by Levin and Smith (1991) and adopted by Paarsch (1992) (among others) for his empirical studies. In his model, Thiel (1988) imposes several restrictions on the random variables C and A_i to simplify equilibrium strategies of bidders. He first assumes that the bidders have a diffuse prior about the true value of the item. This assumption is equivalently stated by taking a prior density function $f(c)$ assumed to be uniform on $(\bar{c}, \underline{c})$. He also assumes that the estimations A_j of the true value c are independently and identically distributed with a variance σ^2, and are unbiased estimates of c, i.e.:

$$c = E[A_j | C=c]$$

The last assumption is that the estimation errors, a_i-c, are statistically independent of the true value for c:

$$F'(a_i-c | c) = f(a_i-c | c) = f(a_i-c)$$

Under these restrictions, equilibrium strategy reduces to (cf. Levin and Smith 1991):

$$b^*(a_i) = a_i - \frac{1+K_2}{K_1} + \beta \exp(-K_1 a_i) \tag{6}$$

where :

$$K_1 = \frac{n(n-1)}{\sigma} \int F(u)^{n-2} f(u)^2 du$$

$$K_2 = n(n-1) \int u F(u)^{n-2} f(u)^2 du \tag{7}$$

$$u = \frac{a_i - \bar{c}}{\sigma}$$

Econometrics of Auctions

Whatever relevant paradigm applies (i.e., whatever the relevant optimal bid function), the parameters of interest of a specified model can be estimated by two categories of approaches: maximum likelihood methods (ML) and nonlinear least squares methods (NLS). Specification of the optimal strategy leads to the formulation of the probability distribution function (hereafter pdf) of the winning bids which is contingent on the pdf of private values or the private estimation of *ex post* value (signals). The observation of these winning bids for a set of independent auctions enables us to determine the likelihood of the model and to derive estimations of the parameters by maximization. On the other hand, NLS methods consist of the minimization of the distance between observed winning bids and the mathematical expectation of these bids, which are functions of the theoretically optimal strategy. One can apply these two kinds of methods to the case of independently distributed private values and to the case of common values.

Maximum Likelihood Methods

Private values paradigm: If we denote by the subscript 1 a specific auction, and assuming that there are n bidders who have the same density function f_1 of private value, the pdf of the winning bid is given by:

$$h_1^P(b_1^w) = n f_1(b^{-1}(b_1^w))[F_1(b^{-1}(b_1^w))]^{n-1} \frac{db^{-1}(b)}{db}$$

Where F_1 is the cumulative distribution function (hereafter cdf) of the private values.

The optimal strategy of a bidder is given by (3), and if we write:

$$\frac{db^{-1}(b)}{db} = \frac{1}{db^{-1}(b)/dv}$$

We obtain:

$$\frac{db(v)}{dv} = (n-1) \frac{f_1(v)}{F_1(v)} \int_v^{v_i} [F_1(x)]^{n-1} dx = (n-1) \frac{f_1(v)}{F_1(v)}(v-b)$$

Hence:

$$h_1^P(b_1^w) = \frac{n}{n-1} \frac{F_1[b^{-1}(b_1^w)]^n}{b^{-1}(b_1^w) - b_1^w} \tag{8}$$

The likelihood function, given the observed winning bids, is:

$$L = \prod_l h_1^P(b_1^w)$$

An explicit form of the pdf of the private value is unlikely to be obtained. Consequently, we are generally unable to give a closed form of the pdf . A solution (Paarsch 1992) is to choose a particular cdf F that gives a functional form to b^{-1} (.).

Common Value Paradigm: The optimal equilibrium strategy is given by (6) and (7). For a particular auction l, the pdf of the winner's signal $a_1^w = a_1(b_1^w)$ is:

$$\tilde{f}_1(a_1^w \mid c) = nF_1(a_1^w \mid c)^{n-1} f_1(a_1^w \mid c)$$

where $f_l(a \mid c)$ is the density of the signal a, given the actual common value c, and $F_l(a \mid c)$ the respective cdf.

Consequently, the density function of the winning bid is :

$$h_1^c(b_1^w) = \frac{\tilde{f}(b^{-1}(b_1^w) \mid c)}{b'(b^{-1}(b_1^w))}$$

Assuming that signals are normally distributed with mean c and variance σ^2, the conditional density is given by:

$$\tilde{f}_1(a_1^w \mid c) = \frac{n}{\sigma} \Phi(\frac{a_1^w - c}{\sigma})^{n-1} \phi(\frac{a_1^w - c}{\sigma})$$

Where Φ and ϕ are the respective cdf and pdf of the unit Gaussian random variable. The optimal bid function is given by:

$$b(a) = a - \alpha_n \sigma + \beta \exp(-a\xi_{1:n}/\sigma)$$

with:

$$\alpha_n = \frac{\int_{-\infty}^{+\infty} nu^2 \Phi(u)^{n-1} \phi(u) du}{\int_{-\infty}^{+\infty} nu\Phi(u)^{n-1} \phi(u) du} = \frac{\int_{-\infty}^{+\infty} nu^2 \Phi(u)^{n-1} \phi(u) du}{\xi_{1:n}}$$

Where $\xi_{1:n}$ is the expectation of the standardized value of the winner,

$$y_1 = \frac{a_1^w - c}{\sigma}, \text{ and}$$

$$b'(a) = 1 - (\beta\xi_{1:n}/\sigma)\exp(-a\xi_{1:n}/\sigma)$$

So, we can obtain the density function of the winning bid :

$$h_1^c(b_1^w) = \frac{n\,\Phi(\dfrac{a_1^w - c}{\sigma})^{n-1}\,\phi(\dfrac{a_1^w - c}{\sigma})}{\sigma[\,1 - (\beta\xi_{1:n}/\sigma)\exp(-a_1^w\xi_{1:n}/\sigma)\,]} \qquad (9)$$

Where a_1^w is the solution of:

$$b_1^w = a_1^w - \alpha_n\sigma + \beta\exp(-a_1^w\xi_{1:n}/\sigma)$$

Consequently, assuming Gaussian estimates a_1^w, the likelihood of the model is given for the observed winning bids by:

$$L = \prod_1 h_1^c(b_1^w) \qquad (10)$$

Nonlinear Least Squares Methods

The NLS methods are based on the minimization of:

$$Q(\theta) = \sum_{l=1}^{L}\,(b_1^w - E(b_1^w))^2 \qquad (11)$$

where θ is the vector of parameters. We will only derive the method within the independently distributed private value paradigm, but adaptation of the method to the independently (but conditional to the same value c) distributed signals case is straigthforward.

Assuming that private values of the n bidders follow the same distribution function, F_1, for each auction l, with:

$$F_1(v_i) = F(v_i, \alpha_1)$$
$$\alpha_1 = g(z_1, \theta)$$

where z_1 is the specific vector of characteristics of each auction l, the minimization of the criterion function requires the knowledge of $E(b_1^w)$, which generally has no closed form. The private value paradigm permits us to simplify this expression. In this case, as proposed by Laffont, Ossard and Vuong (1991), one can apply the Revenue-Equivalence Theorem (Myerson 1981), which states that the seller's expected revenue is the same whether a first or second price auction is designed. Thus, one can write:

$$b_1 = E(v_{(n-1)} \mid v_{(n)})$$

with $v_{(n)}$ being the highest private value and $v_{(n-1)}$ the second one. Consequently:

$$E(b^w) = E(v_{(n-1)})$$

$$= \int \dots \int v_{(n-1)} f(v_{(1)}) \dots f(v_{(n)}) dv_{(1)} \dots dv_{(n)}$$

$$= \int m(m-1) v F(v)^{m-2} f(v)(1-F(v)) du$$

In this expression, $E(b^w)$ does not contain the optimal strategy. (This simplification is not possible within the common value paradigm, since the equivalence revenue theorem does not hold in this case.) Estimation still requires numerical integration over the pdf of the private values. A solution is to estimate it numerically at each stage of the Gauss-Newton estimation procedure, which involves a great number of numerical integrations. An alternative solution (following Pakes and Pollard 1989; Gourieroux and Monfort 1990; or Laffont, Ossard, and Vuong 1991) is to replace the theoretical moment by a simulated empirical estimation to avoid such computations. As noted by these authors, this substitution generally presents the inconvenience of providing only biased estimators, but that can be avoided in the case of NLS. Thus, the Laffont *et al.* (1991) method consists of the substitution of both the first and second order theoretical moments of b^w by simulated empirical estimations. (For an alternative procedure, which involves two sets of simulations, see Gourieroux and Monfort 1990).

Expanding (11):

$$Q(\theta) = \sum_{l=1}^{L} (b_l^{w2} + E(b_l^w)^2 - 2b_l^w E(b_l^w))$$

For each auction l ($l = 1,\dots,L$), we generate S independent draws, each of size n. For each draw s ($s = 1,\dots,S$), we rearrange the draws in increasing order $(v_{(1)l}^s,\dots,v_{(n-1)l}^s,v_{(n)l}^s)$. If we denote $X_{sl}(\theta) = v_{(n-1)l}^s$, the empirical mean $\bar{X}_1(\theta)$ over the simulations is an estimator of $E(b_1^w)$:

$$\bar{X}_1 = \frac{1}{S} \sum_{s=1}^{S} X_{sl}(\theta) = \frac{1}{S} \sum_{s=1}^{S} v_{(n-1)l}^s$$

By construction of $X_{sl}(\theta)$, we have $E(b_1^w)^2 = E(\bar{X}_1(\theta))^2$. Thus:

$$E(b_1^w)^2 = E(\bar{X}_1(\theta)^2) - var(\bar{X}_1(\theta)) = E(\bar{X}_1(\theta)^2) - \frac{1}{S} var(X_{sl}(\theta))$$
$$\bar{X}_1(\theta)^2 \quad \text{and} \quad \frac{1}{S-1} \sum_{s=1}^{S} (X_{sl}(\theta) - \bar{X}_1(\theta))^2$$

are unbiased estimators of $E(\bar{X}_1(\theta)^2)$ and $\text{var}(X_{sl}(\theta))$. Hence we can define a new function $Q^*(\theta)$, minimization of which gives a consistent estimator of θ:

$$Q^*(\theta) = \sum_{l=1}^{L} \left[(b_l^{w2} + \bar{X}_1(\theta))^2 - \frac{1}{S(S-1)} \sum_{s=1}^{S} (X_{sl}(\theta) - \bar{X}_1(\theta))^2 \right] \quad (12)$$

Assuming that private values v_i are normally distributed, with mean $z_l\beta$ and variance σ^2 ($\theta = (\beta, \sigma)$), the method is, for each auction l ($l = 1,...,L$), to randomly draw S vectors of n values u_{il}^s where u_i is the realization of a unit Gaussian random variable. The simulation $X_{il}(\theta)$ is determined by:

$$X_{il}(\theta) = \sigma u_{(n-1)l}^s + z_l\beta$$

where $u_{(n-1)l}^s$ is the second highest value of the vector $(u_{il}^s)_{i=1,...,n}$.

Application: Tenders of Soft Wheat Intervention Stocks

The framework described above is applied to tenders of soft wheat intervention stocks. Open market tenders are a more complex means of awarding because both quantities and refunds are endogenous.

Weekly tenders are presented in reports issued by the EC Cereals Management Committee. These reports contain information about purchase prices of awarded lots, quantities of each lot, and places where lots are located. Tenders apply to homogeneous lots by prefixing correctives, according to differences in quality and the destination of exports. (That is, to take into account freight costs but also to encourage or discourage exports). The data set covers tenders held between December 1990 and June 1992 (424 awarded lots).

Variables which can explain private or common values can be divided into three sets:

- time variables, which reflect seasonal shifters,

- price variables: posted and actual U.S. export price (i.e. bonuses awarded within EEP are taken into account); internal price, the level of which can act upon the timeliness to export in the world market; exchange rate of European Currency Unit (ECU) versus the U.S. dollar, and
- quantity variables: EC and U.S. exports, and total EC exports awarded the same day.

For each auction l, the variable b_l^w is the winning bid, expressed in ECU. The internal price is expressed in French francs (the parity between the French franc and the ECU is constant over the period of estimation). World prices are expressed in U.S. dollars. Quantities are expressed in million tons. We assume that there are 15 bidding traders for each auction, which seems to be very close to reality.

The Assumption of Independent Private Values

Simulated nonlinear least squares estimation is used within the assumption of normally distributed private values. For each buyer and for each auction, 20 values of a standardized normal variable are drawn.

Inside $Q^*(\theta)$, the theoretical expectation of winning bids is given by the empirical average of $X_{sl}(\theta) = \sigma\, u_{((n-1)l}^s + z_l\beta$, $(l = 1,...,L)$, over S simulations. $u_{(n-1)l}^s$ is the second highest value of $(u_{il}^s)_{i=1,...,n}$, the vector of unit Gaussian random variable realizations. z_l is the vector of explanatory variables.

Time variables are either the day of tendering or season (with monthly dummy variables). None has had a significant influence on winning bids, due to the awarding of several lots on the same day.

Parameter estimates are given in Table 15.1, where n=15 traders and S=20 draws. The exchange rate has a positive effect on bids: *ceteris paribus*, an increasing value for the U.S. dollar involves higher bids, expressed in ECUs. U.S. exports have a negative effect. When these exports increase, European traders have to minimize their purchase price in order to stay competitive in the world market.

The most significant result is the effect of the actual world price. The world price used in our estimation is the U.S. fob price, corrected by an average of bonuses. Except for the exchange rate effect, traders' bids are obviously determined by actual world prices. This is confirmation of the leadership role that U.S. trade plays, and consequently that U.S. export subsidy policy plays. Such a strong result occurs, no matter what estimated specification.

The internal price has a negative effect on winning bids. When this price increases it could be less profitable to export to the world market rather than selling in the European market, so traders can decrease their bids. Otherwise, if the internal price increases, the European Commission would prefer to sell its intervention stocks in the internal market in order to minimize cost. If traders anticipate this behavior they can decrease their bids because there is less competition among them.

Other estimation confirms signs and values of previous effects. But effects of quantities which are tendered for are never significant. Apart from the significant effect of the world price, we have to point out low

TABLE 15.1 Parameter Estimates with Independent Private Values

	σ	4.263
		(0.02)
β	Intercept	9.598
		(0.04)
	US	14.181
	(exchange rate ECU/USD)	(1.00)
	QUS	- 4.769
	(US Exports)	(-0.45)
	WP	0.589
	(actual world price)	(3.43)
	ECP	-0.050
	(EC price)	(-1.11)

Note: t-statistics in parentheses below coefficients.
RMSE = 2.97%.

significance for other effects, even if global fitting is correct (see root mean square percent error, or RMSE). On the one hand, this can be due to the estimator itself, which can induce problems of identifiability between the intercept and the variable:

$$\sum_{s=1}^{S} u_{(n-1)l}^{s}$$

The latter has very little variability, especially when the number S of draws is high. On the other hand, such problems remain with a no-intercept estimation. In this case, the same problem occurs between the previous variable and some of the other shifters which have little variability because of a nonconstant frequency. Several variables keep the same value for many tenders (for instance, all of the ones which take place the same day), so their effect on bids can hardly be identified.

Lastly, the estimate of the private values standard error is of little significance, and it is very small ($\sigma \equiv 4$, while the average of winning bids is about 85 ECUs). It does not directly induce rejection of the assumption of independent private values, but one can presume this assumption not to be accurate in the case of tenders of soft wheat intervention stocks.

The Assumption of Common Value

Here we assume that *ex ante* signals are normally distributed for each auction l, with mean $c_l = z_l \alpha$ and variance σ^2. z_l is a vector of explanatory variables, and α is a vector of parameters. Under this assumption of normality, the optimal winning bid is:

$$b_1^w = a_1^w - \alpha_n \sigma + \beta \exp(-a_1^w \xi_{1:n} / \sigma)$$

where the winner's signal is a_1^w. The vector of parameters to be estimated is $\theta = (\sigma, \beta, \alpha)$.

The nonlinear least squares estimator is used, with simulations to estimate the non-observable variable . Results of the estimation are given in Table 15.2. Explanatory variables are the same as under the assumption of independent private values. Here an estimation without intercept is more satisfactory. All the shifters of the common value have similar effects, compared to the private value estimation. These effects can be interpreted in the same way. The estimated standard error of the signals (σ) is lower than the estimated standard error of the private values, but it is significant here, under the common value assumption. Other estimates are significant too (except β) , which is not the case under the private value assumption.

The estimated coefficient β is infinitesimal, and it is not significantly different from zero. However, an estimation where this coefficient was constrained to be zero induced no significant effects from some explanatory variables, so we kept a specification with β.

The goodness of fit, measured through the root mean square percent error, is the same here as under the assumption of private values (about 3 percent of the winning bid). Lastly, the parameter estimates under the common value assumption are different from the ones under the private value assumption, essentially because they are more significant in the former case.

TABLE 15.2 Parameter Estimates Under Common Value Assumption

	σ	2.344
		(2.22)
	β	- 3 E-219
		(-9.1 E-147)
	U.S.	17.210
	(exchange rate ECU/USD)	(88.1)
	QUS	- 5.333
α	(U.S. Exports)	(-17.8)
	WP	0.618
	(actual world price)	(7174)
	ECP	- 0.0536
	(EC price)	(-9078)

Note: t-statistics in parentheses below coefficients.
RMSE = 3.08%.

Conclusion

In this chapter we have pointed out the importance of refund awarding procedures for EC grain exports. An analysis of these procedures can be drawn from recent developments in auction theory. Econometric methods are derived from this theory within two competing paradigms of the valuation of the object for sale. The empirical results do not yet provide clear conclusions, but some features of the market under study have been highlighted. In particular, the high dependence of European traders' behavior on U.S. export prices is very noticeable.

The derivation of structural approaches of uncompetitive markets like tender procedures has provided important theoretical results that have not yet been completely exploited (and verified) by empirical work. Empirical studies face at least two major difficulties: i) related econometric methods are too recent to be as effective as traditional approaches and ii) structural models directly derived from the theory do not always fit to the particular field to be analyzed.

This research has attempted to cope with some econometric methods within the two major canonical models of auctions in their simplistic formulation (assuming symmetry of auctioneers, independence of their valuations, no collusion among them, as well as myopic behavior). Despite these symplifying assumptions, computational difficulties still remain and have prevented good quality estimates from being obtained. Estimating models which are closely derived from the theory may lead to intractable calculations. The closed form of the likelihood function, whatever paradigm is used, is a typical illustration of difficulties that arise, and important restrictions on the probability density function of the stochastic values have to be made to solve them. Another problem, elegantly solved by Donald and Paarsch (1993) in the particular field they analyze, is caused by the ML method itself. The support of the probability density function may depend upon the parameters of interest that jeopardize the consistency and asymptotic properties of the ML estimators.

The simulation method proposed by Laffont, Ossard, and Vuong (1991) to build NLS estimators presents computational problems. These difficulties seem to be caused by a problem of identifiability between the simulated vector and other exogenous variables. However, the simulated method allows easing of supplementary assumptions that are unavoidable in ML methods, and therefore the computing of the theoretical bids in a closed form. Simulated methods seem to be the most promising ones because they permit the use of complicated objective functions that prevail in structural approaches and are intractable otherwise. Moreover, simulated methods are not confined to NLS methods (see Gourieroux and Monfort 1990 for the variety of applications). For example, Pseudo Maximum Likelihood methods help to avoid the pure ML difficulties.

References

Agra-Europe. 1993. *CAP Monitor.*

Donald, S.G. and H.J. Paarsch. 1993. "Piecewise Pseudo-Maximum Likelihood Estimation in Empirical Models of Auctions." *International Economic Review* 34: 121-148.

Gourieroux, C. and A. Monfort. 1990. "A Simulation Based Inference in Models with Heterogeneity." *Annales d'Economie et Statistique* 20: 69-107.

Graham, D.A., R.C. Marshall and J.F. Richard. 1990. "Differential Payments Within a Bidder Coalition and the Shapley." *American Economic Review* 80: 493-510.

Laffont, J.J., H. Ossard and Q. Vuong. 1991. "Econometrics of First Price Auctions." *IDEI-INRA-USC*, Working Paper No. 91.

Le Roux, Y. 1992. "Perspectives d'exportation de blé tendre et d'orge." *INRA Sciences Sociales* . Paris.

Levin, D. and J.L. Smith. 1991. "Some Evidence on the Winner's Curse: Comment." *American Economic Review* 81: 370-375.

McAfee, R.P. and J. McMillan. 1987. "Auctions and Bidding." *Journal of Economic Literature* 25: 699-738.

Myerson, R.B. 1981. "Optimal Auction Design." *Mathematical Operations Research* 6: 58-73.

ONIC. 1991. *Guide du Commerce Extérieur.* Paris.

Paarsch, H.J. 1992. "Deciding between the Common and Private Value Paradigms in Empirical Models of Auctions." *Journal of Econometrics* 51: 191-215

Pakes, A. and D. Pollard. 1989. "Simulation and the Asymptotics of Optimization Estimators." *Econometrica* 57: 1027-1057.

Thiel, S.E. 1988. "Some Evidence on the Winner's Curse." *American Economic Review* 78: 884-895.

16

Welfare Effects of HFCS Development in the U.S. Sweetener Market

*Vincent Réquillart, Christos Cabolis,
and Eric Giraud-Héraud*

Introduction

The U.S. sweeteners market has been the subject of extensive research, primarily due to the degree and duration of government intervention. Most studies on policy issues can be placed into one of two categories. The first addresses evaluation of the welfare implications of different policy regimes (e.g., Leu, Schmitz, and Knutson (1987); Greer (1992); Schmitz and Christian (1993)). The second concerns the political-economic aspect of sugar policies, that is, how the welfare of market participants affects sugar policies (Lopez 1989).

Government intervention in the U.S. sugar market began around 1789 (Schmitz and Christian 1993). The different instruments used include import quotas and tariffs, target prices and subsidies, loan and purchase programs, and restrictions on cultivated acreage. A goal of these policies has been to restrict sugar supply and to support a domestic price greater, and less volatile, than the world price.

The artificial isolation of the U.S. market has motivated considerable research regarding the net cost of such programs. The result has been a range of estimates, depending on one's assumptions (or rather 'beliefs') about different parameters. Schmitz and Christian (1993) identify the price elasticity of supply and demand, the substitution between sugar and high fructose corn syrup (HFCS), the relevant base world price, and the effect of U.S. sugar policy on the world price of sugar as four such categories. In what follows we will concentrate on the substitutability between sugar and HFCS.

Although most researchers agree that the development of the HFCS industry has benefited from artificially high sugar prices, there is no consensus on how to evaluate the presence of the industry in terms of welfare. Schmitz and Christian (1993) suggest that the emergence of the HFCS industry should be considered a benefit for the economy. Maskus (1993), on the other hand, argues that the development of the HFCS industry should be regarded as a cost to society since it is the outcome of a distorted market.

Abstracting from the above debate, the acceptance of the fact that there exists substitutability between sweeteners can significantly alter the estimates of welfare impacts. Table 16.1, taken from Leu, Schmitz, and Knutson (1987), illustrates that, allowing for substitution, consumer cost decreases considerably, given the quota price premium and the elasticity of the excess supply curve. In order to evaluate welfare implications of different policy instruments, Leu *et al.* (1987) used a general equilibrium approach for the sugar market, taking into consideration the substitutability of corn sweeteners for sugar. Lopez (1989) considered the implications of different policy instruments (import quotas and target prices) on welfare. He developed a political-economic framework to analyze the determinants of the level of U.S. sugar policy instruments based on the economic surpluses of market participants. He used an equilibrium demand for sugar approach, which takes into account the feedback from substitute products.

Lopez and Sepulveda (1985) studied the demand for sugar in the U.S. They found that sugar demand is not significantly affected by HFCS prices. The authors attribute the decrease in sugar consumption to 'trends', which have been higher after the introduction of HFCS. Barros (1992), on the other hand, could not reject the hypothesis that sugar prices affect HFCS consumption growth. He summarizes the relationship between sugar and HFCS as follows:

> "(F)or some uses, they are almost perfect substitutes; for others, they are imperfect substitutes. As a consequence, within some price range, it is reasonable to have both products consumed in the economy."

Our approach employs a differentiated product model to study the derived demand for sweeteners as intermediate inputs and to examine the above results. We analyze the market for sweeteners, taking into consideration the different physical characteristics and technical constraints for their use. The paper is organized as follows: we first provide information on the sweeteners market and on the product

TABLE 16.1 Costs and Benefits of the U.S. Sugar Quota Program, 1983

Quota Price Premium	Elasticity of Excess Supply Curve	Consumer Cost			Net Societal Cost	
		Substitution			Substitution	
		Without	With	Producer Gain	Without	With
(¢/lb.)		————————(million)————————				
7.49	2.37	423	372	169	253	203
	∞	1,856	1,636	598	1,258	1,038
10.70	2.37	926	815	345	580	470
	∞	2,661	2,347	742	1,919	1,605
16.05	2.37	1,769	1,559	578	1,191	981
	∞	4,017	3,546	833	3,184	2,713

Source: Leu et al. (1987).

producers in the market; the last section presents and evaluates the welfare implications of HFCS development.

Characteristics of the Sweetener Market

In this paper the sweetener market consists of sugar (sucrose) and high fructose corn syrup (HFCS). Sugar is produced from cane or beet, and is available in solid (crystal) form. HFCS is produced primarily from corn, is composed of the compounds fructose and glucose, and is available in liquid form. Sugar and HFCS can be differentiated along two dimensions: physical form (i.e., liquid or solid), and sweetness. Table 16.2 provides a summary of the average sweetness level and the physical form for the main sweetener products.

Studies on the substitution of HFCS for sugar have examined primarily the U.S. market, because of the extended use of HFCS in the U.S., especially in the soft drink market. In 1988, the consumption of HFCS in the U.S. was 43.7 per cent of total sweetener consumption. The

TABLE 16.2 Some Characteristics of Sweeteners

Sweetener	Raw material	Physical state	Sweetness
Glucose	Cereals	Liquid	0.5 to 0.7
Isoglucose 42	Cereals	Liquid	0.9
Isoglucose 55	Cereals	Liquid	1.0
Sugar (Sucrose)	Sugar beets	Solid	1.0 (Normalization)

Source: Jacquemin and Guerin (1989).

TABLE 16.3 U.S. Caloric Sweetener Deliveries to End User, 1990

	Consumption of sweeteners in 1990 (10^9 pds eq.su)		
Sectors	HFCS *	Sugar	Total
Confectionery	63	2,553	2,616
Soft drinks	9,914	456	10,370
Baking	840	3,210	4,050
Canning	926	661	1,587
Dairy	468	918	1,386
Other food uses		1,284	1,284
Non Industrial uses	-	6,782	6,782
Total	12,211	15,864	28,075

Source: USDA - Sugar and sweetener outlook.
*Note:** Consumption in HFCS dry expressed in sugar equivalent (HFCS-42 = 0.9; HFCS-55 = 1). We used the following coefficient to convert from commercial basis: 0.77 and 0.71 respectively, for HFCS-55 and HFCS-42. We assume that only the beverage industry uses HFCS-55.

differentiation modeling procedure respectively; the next section presents the model for the sweeteners market and derives the demand for sweeteners; the following section analyzes the strategy of HFCS U.S. Department of Agriculture (USDA 1992) estimates HFCS production for 1992 to be around 8.4 million short tons on a dry basis.

The main consumers of HFCS are the beverage and food industries. Table 16.3 provides the total purchase by type of sweetener for the beverage industry, ice cream and dairy producers, bakery and cereal producers, and the confectionery industry. HFCS dominates the beverage, and to a lesser extent, the canning industry. Sugar, on the other hand, is the dominant sweetener in the baking, dairy and confectionery sectors, and in non-industrial uses.

Notes on Product Differentiation

The term 'product differentiation' was introduced by Chamberlin (1933). Products can be differentiated in a number of dimensions: quality, location, time, consumer information about the existence of a product and so on. A general classification is between vertical and horizontal differentiation. Assume that consumers are faced with a number of goods at the same price. If there exists a consensus in ranking these goods based on a measure index (e.g., a quality index or distance index), the goods are said to be vertically differentiated. If such a ranking is not supported by all consumers, the goods are said to be horizontally differentiated (see

Gabszewicz and Thisse (1986) for a presentation of horizontal and vertical differentiation).

There are two modeling procedures used to capture the demand side of horizontally differentiated products. The first is the non-address approach and stems from Chamberlin's model of monopolistic competition. It has been further developed by Spence (1976) and Dixit and Stiglitz (1977). In the non-address approach, consumer preferences are defined over all possible goods. A crucial characteristic of this approach is that each good competes against all other products. This implies that a ranking of products cannot be deduced by consumer preferences. Depending on whether consumer preferences are assumed homogeneous or heterogeneous, the "representative consumer" model or a variant of Chamberlin's model is used.

The second modeling procedure for horizontally differentiated products is the address approach and follows Hotelling's (1929) spatial model. According to this model, goods are distributed along a particular dimension. Consumers have preferences distributed along the same characteristic dimension. This implies that consumers can be thought of as having different "addresses" or locations along the characteristic space. Therefore, a clear ranking on all goods can be deduced. Furthermore, due to the assumption of indivisibility of the available products, if the ideal good for a consumer is not available, the best substitute from the neighboring products is chosen.

We will use the address approach to analyze the sweetener market. Given the indivisibility of the goods assumption, each consumer can choose only one quality. By relaxing this assumption and accepting product divisibility, a consumer can utilize more than one quality, a fact observed in the sweetener market. Generally, the closer the goods are in the characteristic space, the higher the substitution between them. In the sweetener market, a high distance in the "sweetness" characteristic space benefits sucrose by a "reserved market share". The reason is that consumers are constrained to use the best quality.

Analysis of the Demand for Sweeteners

Eaton and Lipsey (1989) point out that a complete model of product differentiation defines (a) the set of products, (b) the production technology of each good, (c) consumer preferences, and (d) an equilibrium concept. We will analyze all the above points.

Our model can be thought of as a three-level vertical structure. The first level is the up-stream firm (USF) and consists of the sweetener producers. The second level includes manufacturers which use sweeteners as inputs (e.g., soft drink industry, baking industry, and so on). These are the intermediate or down-stream firms (DSF). The third level contains the final consumers (FC) of the DSF's products. The demand for sweeteners depends on the technology of the DSF and on the demand for their final product (i.e., on the demand of the FC). Assume two producers in the first level, a sucrose and an HFCS producer. A model of product differentiation is then applied in the first level, where the HFCS producer derives its best reaction, knowing the price of sugar (due to the tariff) and the demand for sweeteners.

In what follows it is assumed that the decision maker of a food processing unit can isolate the sweetness content when selecting ingredients. The manufacturer will minimize the cost of sweeteners, taking into account technological constraints on the level of sweetness and on the bulk content. The objective function consists of two parts: the cost of the inputs, and the treatment cost required to adjust a sweetener to consumers' preferences. The latter is the additional cost incurred by the DFS if they employ corn sweeteners instead of sugar. The manufacturer's problem can be represented by the following program which provides the quantity of the ith sweetener needed to produce one unit of the final good:

$$
\left\{
\begin{aligned}
&\text{MIN over } \{q_i\} \ \sum_{i=1}^{N} \left[p_i q_i + C_i(q_i) \right] \\
&\sum_{i=1}^{N} k_i q_i \geq km \\
&\sum_{i=1}^{N} q_i \leq m
\end{aligned}
\right. \tag{1}
$$

where p_i and q_i denote the price and the quantity of sweetener i, respectively. In this formulation, k_i represents the level of sweetness of the ith sweetener, while k is the minimum level of sweetness that the final product can have. Here k can be thought of as given exogenously by both the technology of the product and the taste preferences of the final consumer. The maximum quantity of sweetener that the final product can accommodate is represented by m and is given by the technology of the final product. The first constraint represents the sweetness requirement and km can be thought of as the lower bound on the sweetness level for the final product. The second constraint refers to the maximum 'bulk content'.

$C_i(q_i)$ represents the user cost for sweetener i[1]. Since all the corn sweeteners considered in this paper have a similar dry matter content, we assume that the treatment costs $C_i(q_i)$ are similar for all i, and have a constant marginal cost of the form $C_i(q_i) = \theta_i q_i$ with:

$$\theta_i = \begin{cases} \theta & \text{if } i \neq N \\ 0 & \text{if } i = N \end{cases}$$

where $i = \{1, 2, ..., N-1\}$ is the range of corn sweeteners and N denotes the sugar.

The parameter θ is the additional cost due to the utilization of a liquid sweetener. In this model, different 'consumers' of sweeteners (e.g., different firms of the DSF) have different preferences with respect to the 'physical form' of the product. Parameter θ captures the importance of 'physical form' of sweeteners assigned by the different consumers. In this sense it characterizes consumer preferences. The model relies on differentiation in 'taste' with respect to the supplied qualities. This approach is similar to models developed by Mussa and Rosen (1978), Gal-Or (1985), and Champsaur and Rochet (1989)[2].

In order to simplify the exposition of the model we will consider two sweeteners: a corn sweetener ($i=1$) and sugar ($i=2$). Figure 16.1 shows the constraints of the manufacturer's problem (1) in the two-sweetener case. The characteristics in which the sweeteners differ are 'sweetness' and 'physical form'. It is assumed that the sucrose's level of sweetness is always higher than the minimum level required ($k \leq k_2$), since sucrose has the highest level of sweetness (see Table 16.2). We distinguish two cases for HFCS. The first occurs when HFCS's level of sweetness is less than the minimum required level ($k_1 < k$). The optimum solution will be in the segment AB. If the manufacturer can choose only one sweetener it will be sucrose. If this restriction is relaxed, the manufacturer will choose a blend of sugar and HFCS, provided the price of corn sweetener is 'low' relative to the price of sugar. The second case occurs when HFCS's level of sweetness is at least equal to the minimum level required ($k_1 \leq k$). The optimum solution then will be on the segment CD. In this case the manufacturer will use the sweetener which has the lower price-quality ratio. The solution of the manufacturer's problem is provided in Table 16.4. $q_1(\theta)$ and $q_2(\theta)$ indicate the quantities of HFCS and sugar used to produce one unit of final good by a manufacturer who is characterized by θ. The parameter $\hat{\theta}$ denotes the manufacturer who is indifferent between employing any mix along the AB segment (or CD segment depending on the k_1 value). Mathematically, parameter θ is given by the following equation:

$$\theta = \frac{k_1 p_2 - k_2 p_1}{k_2}$$

FIGURE 16.1 The Constraints of the Manufacturer's Problem in the Two Sweetener Case

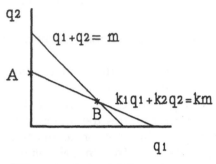

The sweetness of the
HFCS is lower than the
minimum required level
$(k_1 < k)$

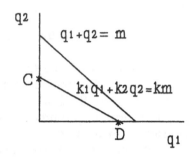

The sweetness of the
HFCS is greater than the
minimum required level
$(k_1 \geq k)$

TABLE 16.4 Optimal Consumption of Sweeteners

	$q_1(\theta)$	$q_2(\theta)$
	$k_1 < k$	
$\theta < \hat\theta$	$\dfrac{k_2 - k}{k_2 - k_1} m$	$\dfrac{k - k_1}{k_2 - k_1} m$
$\theta \geq \hat\theta$	0	$\dfrac{k}{k_2} m$

	$q_1(\theta)$	$q_2(\theta)$
	$k_1 \geq k$	
$\theta < \hat\theta$	$\dfrac{k}{k_1} m$	0
$\theta \geq \hat\theta$	0	$\dfrac{k}{k_2} m$

The DSF's total demand for sweeteners from the USF will depend on the demand for DSF's final product by final consumers. If we denote the demand of the final consumers by $z(p_1,p_2,\theta)$, the demand for sweetener i by the DSF is:

$$D_i(p_1,p_2,k_1,k_2) = \int_{-\infty}^{+\infty} q_i(\theta)z(p_1,p_2,\theta)d\theta \qquad (2)$$

where i=1 for HFCS; i = 2 for Sugar.

In order to derive USF's demand for sweeteners we want to isolate the effect of the differentiated preferences with respect to the sweetness and physical form. To do so we make the following assumption:

$z(p_1,p_2,\theta) = z(\theta)$

To employ this assumption, we appeal to the empirical studies on the elasticity of demand for sweeteners. In general, the aggregate demand for both sugar and HFCS is price inelastic (Schmitz and Christian 1993). In what follows we will assume an inelastic demand and focus our analysis on substitution between sweeteners. Estimates of the own price elasticity of sweeteners for non-industrial use are very low.[3] When sweeteners are used in the agro-food sector (our DSF), they are inputs to production. Since the cost of sweetener is small relative to the price of final goods, a change in the price of sweeteners will have a negligible effect on final demand. Lopez and Sepulveda (1985) estimated industrial demand elasticities for sweeteners to be - 0.15 before the introduction of HFCS-55 and -0.04 afterwards. Therefore, the own price elasticity of sweeteners is likely to be small.

Furthermore, we let the quantity of sweetener i used in one unit of final output relative to the maximum quantity of sweetener that the final product can accommodate, be denoted by $q_i'(\theta)$. That is:

$q_i'(\theta).m(\theta) = q_i(\theta)$

Using the above specification, equation 2 can be written:

$$D_i(p_1,p_2,k_1,k_2) = \int_{-\infty}^{+\infty} q_i'(\theta)m(\theta)z(\theta)d\theta \qquad (3)$$

Assume that the random variable θ has a continuous distribution on the interval (θ^-, θ^+). We also define the quantity:

$$M = \int_{\theta^-}^{\theta^+} q_i'(\theta)m(\theta)z(\theta)d\theta.$$

Since $m(\theta)z(\theta)$ provides the maximum quantity of sweetener that the total output of producer θ can accommodate, M gives the maximum quantity of sweetener for all q's. The function

$$f(\theta) = \frac{m(\theta)z(\theta)}{M}$$

is a (probability) density function and $F(\theta)$ denotes the (cumulative) distribution function.[4] Using the above points, (3) can be written:

$$D_i(p_1,p_2,k_1,k_2) = \int_{\theta^-}^{\theta^+} q_i'(\theta)Mf(\theta)d\theta \tag{4}$$

Since k is given exogenously, the demands for the two sweeteners are as follows:

$$\begin{cases} D_1(p_1,p_2,k_1,k_2) = K_1F(\hat{\theta}) \\ D_2(p_1,p_2,k_1,k_2) = Q_2F(\hat{\theta}) + K_2[1-F(\hat{\theta})] \end{cases} \tag{5}$$

where:

$$K_1 = \begin{cases} \dfrac{k_2 - k}{k_2 - k_1}M & \text{if } k_1 < k \\ \dfrac{k}{k_1}M & \text{if } k_1 \geq k \end{cases} \qquad K_2 = \dfrac{k}{k_2}M \qquad Q_2 = \begin{cases} \dfrac{k - k_1}{k_2 - k_1}M & \text{if } k_1 < k \\ 0 & \text{if } k_1 \geq k \end{cases}$$

In examining (5), we make the following observations:

i. The sweetener choice depends on the price-sweetness ratio. This can be seen by rewriting "$\hat{\theta}$" as follows:

$$\hat{\theta} = k_1\left[\frac{p_2}{k_2} - \frac{p_1}{k_1}\right]$$

ii. The sweetness weighted sum of the demands is constant irrespective of prices. That is, $k_1 D_1 + k_2 D_2 = kM$. This implies that substitution occurs while holding the level of sweetness constant.

iii. Sucrose enjoys a "reserved market" as long as the HFCS's sweetness level is strictly lower than the one required by consumers. Producers of sucrose realize some degree of monopoly power over a range of consumers. This might have happened fifteen years ago when only HFCS-42 was available. The beverage industry utilized at that time a 50-50 percent blend of sugar and HFCS-42. Corn sweetener was priced 20-30 percent below sugar, but further substitution was not feasible due to the "low" sweetness of HFCS-42 (see Cook and Kass 1986).

iv. The demand for sugar varies only if

$$P_2 \in \left[(p_1 + \theta^-)\frac{k_2}{k_1}, (p_1 + \theta^+)\frac{k_2}{k_1} \right]$$

A price change for a sweetener does not necessary imply a change in the final demand for sweeteners.

Best Reply of HFCS Producers

Present U.S. policy consists of a mixture of import quotas and a loan and purchase program. These policies result in sugar prices known by all sweetener producers. This can be thought of as a 'commitment' price. Following this reasoning, we assume that an HFCS producer will choose its price as a strategic variable to derive its best reply to the price of sugar.

In order to simplify the exposition of our model, we consider one HFCS producer who produces a sweetener of quality k_1.[5] Given the price of sugar, the HFCS producer chooses the pair (p_1, k_1) which maximizes its profit. By assuming a constant marginal cost in quantity and quality in the production of the HFCS, the profit function can be written as follows:

$$B_1(k_1, k_2, p_1, p_2) = D_1(k_1, k_2, p_1, p_2)\left[p_1 - C_1(k_1) \right] \tag{6}$$

For a given k_1 we derive the profit maximizing B_1. Giraud-Héraud and Réquillart (1992) have proven that $B_1(k_1, k_2, p_1, p_2)$ is quasiconcave in p_1 when:

$$\psi(\theta) = \theta + \frac{F(\theta)}{f(\theta)}$$

is increasing in θ. The first order condition is given by:

$$\frac{\delta B_1}{\delta p_1} = K_1 F(\theta) - K_1 \left[p_1 - C_1(k_1) \right] f(\theta) \tag{7}$$

The solution of the above equation provides the HFCS producer's best reply, in terms of prices, for a given price of sugar, and is as follows:

$$p_1^{br}(k_1, p_2) = \begin{cases} \dfrac{k_1}{k_2} p_2 - \theta^- & \text{if } U_{1,2} \leq \psi(\theta^-) \\[2mm] \dfrac{k_1}{k_2} p_2 - \psi^{-1}(U_{1,2}) & \text{if } \psi(\theta^-) \leq U_{1,2} \leq \psi(\theta^+) \\[2mm] \dfrac{k_1}{k_2} p_2 - \theta^+ & \text{if } U_{1,2} \geq \psi(\theta^+) \end{cases} \tag{8}$$

where

$$U_{1,2} = k_1 \left(\frac{p_1}{k_2} - \frac{C_1(k_1)}{k_1} \right)$$

Equation (8) suggests that $B_1(P_1^{br})$ is increasing in $k_1 \in (0, k_2)$. This result demonstrates an incentive to produce the higher quality HFCS. It could partly explain the dramatic change in market share captured by HFCS-55 during the last 10 years.

In order to evaluate the best reply of the corn sweetener producers, we need estimates of HFCS demand. Since the parameters of the model cannot be directly observed, a detailed survey was conducted on the main industrial users of sweeteners by Giraud-Héraud, Réquillart, and Tazdaït (1992). The survey provides a ranking for the values of the key parameters of the model and can be found in Table 16.5.

These estimates, however, are based on the form of the probability density $f(\theta)$. The class of probability density function that will be employed is:

$$f_{a,\alpha}(\theta) = \frac{a^{\alpha-1}(\alpha-1)}{(\theta+a)^{\alpha}}$$

where $\theta \in [0, \infty]$ and it is expressed in cents per pound.

We will illustrate the results of our model by using the specific function $f_{6,2}$, which adequately captures the demand for liquid sweeteners in the U.S. market.[6] These results should be considered as qualitative rather than quantitative. Figure 16.2 shows the evolution of the price of sugar, the price of HFCS-55, and the estimated best reply by the HFCS producer in the U.S. market from 1981 to 1991[7]. In general, the HFCS price estimates capture the evolution of observed prices. The big difference between the estimated and the observed HFCS prices in 1982 could be attributed to a different strategy by corn sweetener producers. They can be thought to be charging a lower price in that year in order to capture a larger share of the market. In fact, during 1982, several firms from the beverage industry switched from HFCS-42 to HFCS-55.

Figure 16.3 shows the demand for HFCS when the sugar price varies and the HFCS producer plays its best reply. The lower the initial sugar price, the stronger the effect of a decrease in the price of sugar on the HFCS demand. For a high price of sugar ($0.25 - $0.30 per pound) the effect on HFCS demand is rather negligible. There are two reasons for this. The first concerns strategic behavior. A decrease in the price of sugar implies a decline in the price of HFCS and therefore a drop in "$\hat{\theta}$". The change in "$\hat{\theta}$" will be greater the lower the price of sugar. The second is related to the probability density function. Since $f(\theta)$ is decreasing in θ, a given variation of "$\hat{\theta}$" has a greater effect on demand when "$\hat{\theta}$" is small.

TABLE 16.5 Key Parameters for the Different Sectors of Food Production

| Sectors | Model Parameters[*] | | | Maximum Rate of HFCS [(**)] |
	k	$\theta^{\cdot}$	$\theta^{+}-\theta^{\cdot}$	
Confectionery		4		5
Soft drinks	3	1	1	95
Baking	2	3	2	25
Canning	1	2	2	50
Dairy	2	3	2	25
Other food uses		4		0
Non Industrial uses		4		0
Total				

Source: Réquillart and Giraud-Héraud (1992) and Lord (1990).
Note:[*] *Class of values, 1 represents a low value.*
[**] Maximum theoretical penetration rate of isoglucose (Lord 1990).

FIGURE 16.2 Comparison of Observed and Estimated Sweetener Prices

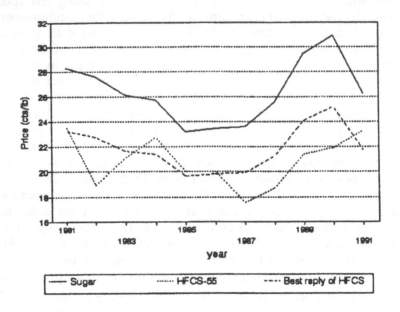

FIGURE 16.3 Best Reply of HFCS to Sugar Price

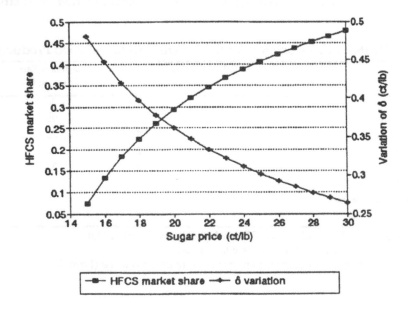

Welfare Implications

In the spirit of the comments made in the introduction, we will focus on the welfare implications of the substitution between sugar and corn sweeteners in the U.S. market. Our objective is to compare the results derived from our model with those of the existing literature in order to evaluate the use of the differentiated product modeling procedure. In this paper we will examine the welfare implications of U.S. sugar policy under the restrictive assumption that the U.S. is a small country (i.e., the excess supply curve of the world is perfectly elastic).

Due to government intervention, the U.S. sugar price (p_2^u) is above the world sugar price (p_2^w). The HFCS producer chooses its best reply in price ($p_1^u = p_1^{br}(k_1, p_2^u)$). We are assuming that the cost-quality ratio of HFCS:

$$\left(\frac{C_1(k_1)}{k_1} \right)$$

is greater than the world sugar price. This implies that, in the absence of the quota policy, HFCS production does not occur and only sugar is available in the U.S. market. When the domestic price is equal to the world price the demand for sugar is denoted by D_2^w and is satisfied by domestic production Q_2^w and imports I_2^w. Under governmental intervention, the demands for sugar and HFCS are given by D_2^u and D_1^u, respectively. In this case, the demand for sugar is satisfied by domestic production Q_2^u and imports I_2^u. We assume that the U.S. is a net importer of sugar. The demand for sweeteners as a function of the prices and the sweetness is denoted by $D_i^j = D_i^j(k_1, k_2, p_1^j, p_2^j)$ where i indicates sweetener (i = 1, 2) and j indicates whether there is government intervention or not (j = u, w). The above can be summarized in the following equations:

$$D_2^u = Q_2^u + I_2^u \tag{9}$$

$$D_2^w = Q_2^w + I_2^w \tag{10}$$

$$k_1 D_1^u + k_2 D_2^u = k D_2^w = kM \tag{11}$$

Equations (12) to (16) provide the gains of the domestic sugar producers (S_s), the HFCS producers (S_h), the exporting countries (S_e), the domestic consumers (S_c), and the net cost (C) of the sugar quota policy:

$$S_s = p_2^u Q_2^u - p_2^w Q_2^w - \int_{Q_2^w}^{Q_2^u} C_2(q)dq \tag{12}$$

$$S_h = \left[p_1^u - C_1(k_1)\right] D_1^u \tag{13}$$

$$S_e = I_2^u \left[p_2^u - p_2^w\right] \tag{14}$$

$$S_c = p_2^{'w} D_2^w - p_2^u D_2^u - p_1^u D_1^u - K_1 \int_{\theta-}^{\theta} \theta\, f(\theta)\, d\theta \tag{15}$$

$$C = S_s + S_h + S_c \tag{16}$$

(15) represents the losses of consumers due to the quota policy. The first term represents total expenditures when the domestic sugar price is equal to the world price. The second and third account for the total expenditures under governmental intervention when both sugar and HFCS are available. The last term denotes the user cost for HFCS.

The net gain in welfare (denoted G_h) due to the introduction of HFCS can be approximated by the difference between the welfare cost when HFCS production is inactive (Cn) and the welfare cost when HFCS production is active (Ch).

$$G_h = C_n - C_h \tag{17}$$

We assume that HFCS production has no direct effect on the welfare of domestic sugar producers in the sense that the sugar price is exogenously given and does not change with the development of HFCS. From (11), (13), (15), (16), and (17)[8], we write the gain due to HFCS production as follows:

$$G_h = \left(D_2^w - D_2^u\right) p_2^u - C_1(k_1) D_1^u - K_1 \int_{\theta-}^{\theta} \theta\, f(\theta)\, d\theta \tag{18}$$

On the right hand side of equation 18, the first term is the gain due to the decrease of sugar consumption when HFCS is available. The second term is the production cost of HFCS. The third term is the additional cost

TABLE 16.6 Welfare Gains from HFCS Production ($ million)

| Sugar price | Density: $f_{6,2}$ | | Density: $f_{5,2}$ | |
| | HFCS cost | | HFCS cost | |
¢/lb	14 ¢/lb	16 ¢/lb	14 ¢/lb	16 ¢/lb
25	1,061	776	1,165	860
20	403	202	453	231

undertaken by consumers when they use HFCS. Since we assume that the HFCS production cost is greater than the world price of sugar, the welfare gain does not depend on the world price. Notice that a change in the price of HFCS has two effects: the first is an adverse effect on the welfare gain (i.e., the welfare gain decreases when the price of HFCS increases); the second is a transfer of surplus between HFCS producers and consumers.

Table 16.6 provides estimates on the welfare gain due to HFCS production for two probability density functions, for different levels of costs of HFCS production, and for different sugar prices. We assume the world sugar price is $0.14 per pound. These results are different from those of Leu, Schmitz, and Knutson (1987) (see Table 16.1). For example, with a sugar price premium of $0.11 per pound (i.e. a U.S. price for sugar around $0.25 per pound in our model) and a perfectly elastic excess supply curve, they found a net gain of about $300 million. Estimates provided in Table 16.6 vary between $750 and $1,150 million. Our results suggest that the cost of the US sugar policy is not as high as it was first assumed.

Conclusion

We have proposed a product differentiation approach in order to analyze sweetener markets in the U.S. Given government intervention, the price of sugar can be seen as a "commitment" price. Following this reasoning, we modeled HFCS producers as choosing their price as a strategic variable. We employed a model of product differentiation where intermediate firms can utilize more than one quality and where products are differentiated according to two parameters, namely sweetness and physical form. We modeled the derived demand for sweetener as intermediate input rather than as a consumption good. This approach could be used to analyze the demand for many agricultural products which are often intermediate inputs for the agro-food system.

We applied this model to the U.S. sweetener market in order to examine the welfare implications of U.S. sugar policy. Our objective was to compare the results given by our model with those in the existing literature. Employing the restrictive assumption that the U.S. faces a perfectly elastic excess supply curve, we show that the cost of U.S. sugar policy is not as high as other studies have suggested. As shown by the model, the best reply profit function of cereal sweeteners producers increases with quality. This means that sugar policy has caused them to alter characteristics of cereals sweeteners and to make them a closer substitute to sugar.

There are several ways in which the above research can be extended. With respect to the modeling procedure, a comparison between the stylized facts of the sweeteners market and the results of our model will show whether the Bertrand-Nash model contributes to our understanding of this market. Additionally, in order to simplify the analysis, we considered only one quality of HFCS. An apparent question is whether the results of our model hold if we increase the number of corn sweeteners under consideration.

With respect to welfare implications, a natural extension is to examine the results of our model under the 'large country' assumption. In this case, U.S. sugar policy affects the world price and therefore the development of the HFCS will be seen from a global perspective.

Notes

1. For example, in the soft drink industry, this cost is low because the utilization of a liquid sweetener is easy. In the jam industry, on the other hand, an additional cost appears because when a liquid sweetener is used, the water added to it has to be evaporated. In some sectors where it is impossible to use a liquid sweetener, i.e., chocolate industry, the additional cost is very high.

2. In contrast, traditional models of vertical differentiation by Gabszewicz and Thisse (1979) and Shaked and Sutton (1982) use income to distinguish the consumers.

3. According to Huang (1985), the own price elasticity is -0.05 and according to Lopez (1989) it is -0.16.

4. The distribution function takes the following values:

$$F(\theta) = 0 \text{ if } \theta < \theta^-; F(\theta) = \int_{\theta^-}^{\theta} f(x)\,dx \text{ if } \theta^- \leq \theta < \theta^+; F(\theta) = 1 \text{ if } \theta \geq \theta^+.$$

5. In the US market, HFCS-55 represented only 25% of total corn sweeteners sales in 1980, while in 1992 this number increased to 60%.

6. This is because $F_{6,2}(3) = 1/3$ and $F_{6,2}(12) = 2/3$ which can be interpreted as follows: 1/3 of the total sweetener consumption corresponds to low values of

q ($\theta<3$, represents mainly the beverage industry) and 1/3 corresponds to high value of θ ($\theta>12$, represents the confectionery industry and non industrial uses).

7. For $f_{a,2}(\theta)$, the best reply is given by

$$p_1^{br} = \frac{k_1}{k_2} p_2 + a - [(\frac{k_1}{k_2}p_2 - C_1(k_1) + a)\, a]^{1/2}.$$

The marginal cost of HFCS-55 production is taken equal to \$0.14 per pound. This estimate is derived from Cook and Cass (1986).

8. If HFCS production does not exist, Sc becomes $D_2^w [p_2^w - p_2^y]$ due to the assumption of an inelastic sweeteners demand.

References

Barros, A.R. 1992. "Sugar Prices and HFCS Consumption in the United States." *Journal of Agricultural Economics* 43: 64-73.

Chamberlin, E. 1933. *The Theory of Monopolistic Competition*. Cambridge, MA: Harvard University Press.

Champsaur, P. and J.C. Rochet. 1989. "Multiproduct Duopolists." *Econometrica* 57: 533-537.

Cook, B.M. and J.S. Kass. 1986. "The Corn Sweetener Industry." Salomon Brothers Inc., Stock Research.

Dixit, A.K. and J. Stiglitz. 1977. "Monopolistic Competition and Optimum Product Diversity." *American Economic Review* 67: 297-308.

Eaton, C.B. and R.C. Lipsey. 1989. "Product Differentiation," in R. Schmalensee and R.D. Willig, eds., *Handbook of Industrial Organization*. Amsterdam: North Holland.

Gabszewicz, J.J. and J.F. Thisse. 1979. "Price Competition, Quality and Income Disparities." *Journal of Economic Disparities* 20: 340-359.

Gabszewicz, J.J. and J.F. Thisse. 1986. "On the Nature of Competition with Differentiated Products." *Economic Journal* 96: 160-172.

Gal-Or, E. 1985. "Differentiated Industries Without Entry Barriers." *Journal of Economic Theory* 37: 310-339.

Giraud-Héraud, E. and V. Réquillart. 1992. "Potential Competition with Vertical Product Differentiation: The Example of the EC Sweetener Market." INRA-ESR Grignon, Working Paper No. 92.06.

Giraud-Héraud, E., V. Réquillart, and T. Tazdaït. 1991. "Sweetener Market: An Analysis of Competition Between Glucose, HFCS, and Sugar." INRA-ESR, Notes and Documents 40.

Greer, Thomas V. 1992. "The Impact of Trade Liberalization on the World Sweetener Industry." Unpublished Ph.D. Thesis. West Lafayette, IN: Purdue University.

Hotelling, H. 1929. "Stability in Competition." *Economic Journal* 39: 41-57.

Huang, K.S. 1985. "US Demand for Food: A Complete System of Price and Income Effects." Washington, DC: USDA, ERS, Technical Bulletin No. 1714.

Jacquemin, C. and B. Guerin. 1989. "Sweeteners, Technological Value, and Utilization." *Actualités Scientifiques et Techniques en Industries Agro-alimentaires.* n°43, CDIUPA.

Leu, Gwo-Jiun M., A. Schmitz, and R. D. Knutson. 1987. "Gains and Losses of Sugar Program Policy Options." *American Journal of Agricultural Economics* 69: 591-602.

Lopez, R.A. 1989. "Political Economy of US Sugar Policies." *American Journal Agricultural Economics* 71: 20-31.

Lopez, R.A. and J.L. Sepulveda. 1985. "Changes in the US Demand of Sugar and Implications for Import Policies." *Northeast Journal of Agricultural Resources and Economics* 14: 177-182.

Lord, R. 1990. "Canadian Sugar and HFCS Industries and US Trade." *Sugar and Sweeteners.* Washington, DC: USDA, ERS.

Maskus, K.E. 1993. "Comment on A. Schmitz, and D. Christian: The Economics and Politics of the U.S. Sugar Program," in S. V. Marks and K. E. Maskus, eds., *World Sugar Policies at a Crossroad.* Ann Arbor, MI: University of Michigan Press.

Mussa, M. and S. Rosen. 1978. "Monopoly and Product Quality." *Journal of Economic Theory* 18: 301-317.

Réquillart, V. and E. Giraud-Héraud. 1992. "Product Differentiation in Sweeteners Market: An Analysis Model." *Cahiers d'Economie et Sociologie Rurales* 23: 6-34.

Schmitz, A. and D. Christian. 1993. "The Economics and Politics of the U.S. Sugar Program," in Steven V. Marks and Keith E. Maskus, eds., *World Sugar Policies at a Crossroad.* University of Michigan Press, 49-78.

Shaked, A. and J. Sutton. 1982. "Relaxing Price Competition Through Product Differentiation." *Review of Economic Studies* 44: 3-13.

Spence, M. 1976. "Product Selection, Fixed Costs, and Optimum Product Diversity." *Review of Economic Studies* 43: 217-235.

USDA. "Sugar and Sweeteners: Situation and Outlook Yearbook." Washington, DC: USDA. Various Issues.

17

Virtual Decisions Under Imperfect Competition: The International Coffee Agreement

Jay S. Coggins

Introduction

From 1963 to 1989 the world market in coffee operated in more or less orderly fashion under a remarkable international agreement amongst importing and exporting nations. Under the International Coffee Agreement (henceforth the ICA) each member exporting country is annually assigned a quota that places a limit upon the level of its exports to member importing countries. Akin in certain respects to an international cartel, the ICA is, nevertheless, almost of its own kind in international trade. It is the exemplar among the class of International Commodity Agreements, of which there have been five noteworthy examples in recent history. The other four—for cocoa, sugar, natural rubber, and tin—enjoyed only modest success at best. The five have in common their inclusion of both exporting and importing countries as members. The coffee agreement alone managed to achieve anything like its original promise, functioning with one or two interruptions for over a quarter century.

A cartel, whether among producing nations or among producing members of a domestic industry, has as its aim the drawing together of producers to limit output, raise prices, and thereby secure enhanced profits for members. Two abiding themes unify the study of cartels. One is that perpetrators and victims may be easily distinguished. If it is successful, producers gain and consumers lose. The other is that a cartel, in its pure form, is stubbornly unstable (Stigler 1964; Osborne 1976; and Donsimoni, Economides, and Polemarchakis 1986). From the viewpoint of a planner whose aim is to maximize aggregate profits for member firms, there is (for the garden variety version) an unambiguous "best" outcome, perhaps consisting of a vector of production levels for each member. The agents who populate the model, however, face a strong incentive to defect from this outcome. Each one, assuming that others will adhere to the cartel agreement, can gain by defecting and producing more than its allocation.

This simple version of the cartel problem is well understood, but there is an extensive literature treating richer and more realistic models. At bottom, the problem for a cartel is one of information gathering and transmission, for if it were costless to monitor and sanction the other members of the cartel one might expect an optimal outcome to be sustainable. The literature along these lines is vast; examples include studies by Porter (1983) and Green and Porter (1984). The hallmark of this literature is a focus on strategic interaction between actors; it often relies upon game theoretic tools for its technical foundation. Studies of international cartels and imperfect international competition, among them studies by Baldwin and Clarke (1987), Harrison and Ruström (1991), likewise emphasize the strategic nature of the cartel problem.

In their recent paper Karp and Perloff (1993) devise a linear-quadratic dynamic oligopoly model of Brazil and Colombia, the world's two largest coffee exporters. Their model captures explicitly the dynamic features of the coffee market, and is designed to assess the degree of competitiveness between the two countries, while treating other exporters as a competitive fringe. It does not account for the influence of importing countries in setting export quotas for Brazil and Colombia.

Using data from 1961/62 through 1983/84, they find that the two countries behave competitively. That is, according to their results the steady-state export levels of the two countries are very close to those that would obtain if the countries were price takers. Though their treatment of the dynamics of the market is more realistic than the static model of this chapter, I take explicit account of the influence of importing countries on the export quota decisions by the ICA. These are decisions that both Brazil and Colombia appear to have abided by, and whose inclusion here mark the primary distinction between the two analyses.

A commodity agreement is a wholly different object than a cartel, and in some respects more puzzling. This claim may perhaps best be supported by looking at what I have called the two themes of the study of cartels. First, a commodity agreement expressly blurs the lines between winners and losers: producers and consumers alike belong. If the interests of cartel members are opposed, then the interests of members of a commodity agreement are much more so--diametrically even. What is good for producers, on the whole, is bad for consumers, and vice versa. Second, the problem of maintaining the collective decision--the *agreement*--may not be less difficult, but is certainly different for a commodity agreement than for a cartel. As we shall see, there is a built-in enforcement mechanism. A producer simply cannot sell more than its quota unless it acts in concert with another member of the agreement, the exporter to whom it sells.

A number of surveys of the ICA have been written, including Fisher (1972), Marshall (1983), Pieterse and Silvis (1988), and Wrigley (1988). Before 1960 the world coffee market was characterized by volatile prices and supply conditions. The ICA was formed in 1963 in response to this volatility. It was headquartered in London and administered by the International Coffee Organization (henceforth, the ICO). The U.S., the world's largest importer of coffee, played a leading role in the formation of the Agreement. The Agreement was renewed, with some revisions, in 1967, 1976, and 1983 (Gordon-Ashworth 1984). It finally broke down when the attempt to produce yet another version failed in July of 1989. Since that time the ICA has been in effect, but without "economic provisions." The Agreement expired on September 30, 1993.

The Agreement has two primary goals: to achieve a balance between world supply and demand; and to moderate price fluctuations. It is, however, a pure export control agreement, concerned first and foremost with controlling export quantities. (The cocoa and natural rubber agreements are buffer stock agreements -- see Gilbert 1984.) Evidently the developed-country member importers also sought a subsidiary goal--that of providing a certain level of export revenue to developing country exporters.

Being an export control agreement, the central mechanism of the ICA is the setting of quotas or market shares for exporters. Numerous additional features and details are also included, among them a provision for modifying the quotas in response to extreme movements in a global indicator price. Exporters are required to export their yearly quota amounts evenly throughout the year, so as to avoid seasonal surpluses and shortages.

Each year the ICO meets in London in September to revise the quota amounts in response to current and expected market conditions. The meeting is attended by all importing and exporting members, who constitute the Coffee Council. With the exception of rules changes (which require a two-thirds majority) their decisions are produced by majority vote.

The voting is itself weighted according to the size of each country's historical import quantities (for importers) and export quotas (for exporters), but votes are divided equally between the importing and exporting groups. Specifically, a total of 2000 votes are distributed among the member countries, 1000 to exporters and 1000 to importers. The collective will of the Agreement is compiled into a final annual decision--the selection of quota levels for member exporters--at this meeting, where a draft proposal is produced by a handful of the larger members, and then submitted to a final vote by the entire council.

The central elements of the decision-making apparatus of the ICA are now before us. It is my purpose in this chapter to make sense of them formally. That is, I shall devise a model that, given information on the preferences of members, produces as an outcome a vector of export quotas.[1] I seek a criterion function for the ICA--an objective function-- that "rationalizes" its behavior. I wish to devise a model of collective decision-making such that the data we observe would have been generated if agents behaved according to the model. My search can be judged successful if, when available information on the primitives of the model (preferences, agents, and alternatives) is employed, the ICA's observed behavior yields a maximum to the criterion function on offer. In order to make the problem manageable, the 40 exporting members and the 25 importing members are aggregated into 7 groups, and published elasticity data are used to derive the demand and cost information that determines their preferences over possible quota vectors. I then specify a criterion function, which I call a welfare function, that consists of a weighted sum of logged profits (for exporters) and consumer surpluses (for importers).

The demise of the ICA in 1989 is most fortunate for this study. With the empirical calculations that appear later in this chapter I am able to project, using only information that was available before 1989, what the coffee market would look like in the absence of the ICA, if all countries were to take the world price as given. That is, I calculate the equilibrium price and quantity traded in a competitive market. These predictions can be readily compared to the events that have transpired in the four years since 1989. This comparison is perhaps the leading finding of the paper, for my model performs quite well in projecting these recent events. A second finding, also positive in nature, is that the collective welfare function that I specify for the ICA does indeed rationalize the outcome that we see in the data.

A Model of Aggregate Decision-Making

The General Implementation Problem[2]

Though the ideal notion of "optimum" is well-defined when one treats the problem of an individual producer or consumer, it is not at all clear what criterion guides decision-making by a body of several agents. From the social choice literature it is known that the leading regularity of social decision problems is the impossibility of devising a well-behaved optimization program for the group. If a mild set of conditions on individual preferences and the decision rule are satisfied, then aggregate preferences are *dictatorial* (Arrow 1963) or *manipulable* in that members can gain by lying about their individual preferences (Gibbard 1973; Satterthwaite 1975).

What might be done to rescue an optimizing framework for a collective body like the ICA? The remedy I employ is found in the theory of implementation with complete information. (For a recent review of related literature see Moore 1992.) The idea of implementation is to devise a game, with members of the social choice setting as players, whose equilibrium coincides with the outcome of a given social choice rule. Maskin (1977) first studied the problem of exact implementation in Nash equilibrium. Matsushima (1988) and Abreu and Sen (1991) recently generalized Maskin's result using the notion of "virtual implementation," which involves introducing an element of randomness into the implementing game. For present purposes the virtue of Abreu and Sen's (1991) approach is that almost any social choice rule can be implemented virtually, though exact implementation is difficult.

The general collective decision problem considered here includes four basic elements: a set of (feasible) outcomes; a collection of decision-makers; these decision-makers' preferences over outcomes; and a social choice correspondence f. Let the group of agents--the *council*--be $I = \{1,...,N\}$. Members of the council are of two types: there are n_1 exporters and $n_2 = (N - n_1)$ importers. The council's task is to select an *outcome vector* $q = (q_1,...q_{n_1})$ of quota levels, where q_i denotes the level of marketings that exporter i will be permitted to sell on the world market. The set of feasible outcomes $Q \subset R_+^{n_1}$ is a countable, dense subset of the n_1-dimensional positive orthant. Let $\mathfrak{L}$ denote the set of lotteries over Q. That is, an element of $\mathfrak{L}$ is a vector of probabilities, one for each of the elements of Q.

Member i possesses preference ordering over Q, $R_i(\Theta) \subset Q \times Q$, where Θ_i indexes i's preferences over Q.[3] Given any two elements q, r ε Q, $qR_i(\Theta)r$ means that i prefers q to r. Let $\Theta = (\Theta_1,...\Theta_N)$ be an element of

Σ, the set of *admissible* preferences. Thus, Θ, without the i index, is a *profile*: it contains information on each member's preferences. The strong assumption of complete information requires that each member knows the preferences of every member with certainty.

A *social choice correspondence* is a rule $f : \Sigma \Rightarrow Q$ that associates with each profile a non-empty subset of Q.[4] A *mechanism* G assigns to each agent a strategy set S_i, and provides a rule for associating to each vector of strategies an outcome (payoff) vector. G is also called a *game form*. A game form is distinguished from a game in that a game form must provide the rules for strategic interaction conditional on a (possibly unknown) Θ. Let $G = (S,g)$, and let $g : S \Rightarrow Q$ be a *payoff function* that maps the strategies of players into elements of Q. We write $G = (S,g)$. Let $s_{-i} = (s_1,...,s_{i-1}, s_{i+1},...,s_N)$ be the (N-1)-dimensional vector obtained by deleting the ith element of s. In this paper the equilibrium concept under study is the Nash equilibrium. A Nash equilibrium in G is a vector of strategies at which no agent can gain by unilaterally adopting a different strategy. A *Nash equilibrium* for G is a strategy vector $s^* \in S$, denoted $N(G,\Theta)$, such that for each $i \in I$, for each $\Theta \in \Sigma$, $g(s^*)R_i(\Theta)g(s_i,s_{-i}^*)$ for each $s_i \in S_i$.

The problem of implementation concerns the logical relationship between f and a given game form G. Say that G *implements* f if whenever a council with profile Θ plays the game G, the equilibrium of the game and the outcome of $f(\Theta)$ coincide. Whether using f or G, the problem is to map preferences into outcomes. The social choice correspondence f is *implementable* in Nash equilibrium if there is a game form G such that for every $\Theta \in \Sigma$, $N(G,\Theta) = f(\Theta)$. If one adheres to the setup of Abreu and Sen (1991), then almost any f can be implemented *virtually*—that is, the equilibrium outcome of the implementing game can be guaranteed to be arbitrarily close to the outcome under f.

Given this setup, the model is complete upon specification of a suitable social choice correspondence. This correspondence, it will be recalled, maps a vector of preferences into a vector that assigns a quota amount to each exporter. I now turn to a development of the virtual implementation setup for an international commodity agreement.

Implementing a Commodity Agreement Virtually

As in the implementation problem, the implementing game for the ICA requires specifying a set of agents, a set of outcomes, each agent's preferences over these outcomes, and a social choice correspondence. The set of agents consists of all exporting and importing members of the Agreement. Let I denote the council, consisting of n_1 exporters and n_2 importers. An exporter will be denoted i, and an importer will be denoted j. The set of outcomes Q is a subset of $R_+^{n_1}$. An element of this outcome space is $q = q_1,...,q_{n_1}$, specifying the export quota for each exporting member. Because the coffee price will influence the payoffs to importers and exporters alike, and because the price depends on q, preferences of the two groups are interrelated. The general formulation of these preferences is presented here, and in the next section they are calculated for the problem at hand.

Exporter i exports q_i, and achieves profit:

$$\pi_i(q_i;q_{-i},c_i,d_i) = q_i \cdot P(q) - C(q_i; q_{-i},c_i,d_i),$$

where P(q) is the world coffee price depending upon the entire quota vector (via aggregate import demand), and where C_i is i's cost function, with parameter vector $\Theta_i = (c_i,d_i)$. By assumption the cost function is quadratic in q_i:

$$C_i(q_i;q_{-i},\delta_i) = c_iq_i + d_iq_i^2.$$

An exporter's preferences are given by its profit function. That is, for any two quota vectors q and r we write:

$$qR_i(\Theta_i) r \text{ if and only if } \pi_i(q,c_i,d_i) > \pi_i(r,c_i,d_i). \tag{1}$$

Note that exporter i has a linear marginal cost function, given by:

$$MC_i(q_i;c_i,d_i) = c_i + 2d_iq_i.$$

Importer j has linear import demand function $Q_j = (a_j - P)/b_j$.[5] World import demand is the piecewise linear function given by the horizontal summation of the Q_j. Denote the inverse world import demand by $P(Q_1,...Q_{n_2})$. Abusing notation slightly (and using the fact that $\Sigma_i q_i = \Sigma_j Q_j$), aggregate import demand will sometimes be denoted P(q). The pair of parameters a_j and b_j index j's preferences: $\Theta_j = (a_j,b_j)$. It is assumed that importer j's preferences are determined from its consumer's surplus at a given price P:

$$CS_j(P(q);a_j,b_j) = \frac{1}{b_j} \int_{P(q)}^{a_j} (a_j - z)dz = \frac{(a_j - P(q))^2}{2b_j}.$$

For two outcome (quota) vectors q and r we write:

$$qR_j(\Theta_j)r \iff CS_j(P(q),a_j,b_j) > CS_j(P(r),a_j,b_j). \tag{2}$$

A profile of preferences over Q—one preference ordering for each exporter and each importer—is given by:

$$\Theta = (c_1, d_1 \ldots c_{n1}, d_{n1}; a_1, b_1, \ldots, a_{n2}, b_{n2}).$$

All but one of the four primitive elements of the implementing scheme are now in place. It only remains to specify a social choice correspondence for combining preferences into a quota outcome. It is assumed that this function takes the form of an asymmetric Nash bargaining criterion, with threat point denoted z_i and z_j in the case of an exporter and an importer respectively. The threat point in a Nash bargaining game is the payoff that each player would obtain if the bargaining process were to break down. It is a catastrophic outcome--the worst that can happen. I assume that the exporting countries have threat points of zero, which corresponds to a coffee price at which profits are driven to zero for all exporters. Importers, on the other hand, have threat points given by the consumers surplus achieved at the highest price that has been experienced in recent years. From their viewpoint, this is the worst possible outcome. In 1977, the New York composite coffee price rose to $3.05 per pound, which is equivalent to $576.72 per 60kg bag in constant 1985 dollars. Importer j's threat point is simply the value of its consumer surplus at $576.72.

Each exporting and each importing member is assigned a weight, and these weights enter the Nash bargaining criterion as exponents. These weights are such that $\Sigma_i \alpha_i = 1$ and $\Sigma_j \beta_j = 1$.[6] The social choice correspondence $f(\Theta)$ is given by:

$$f(\Theta) = \underset{q \in Q}{\arg\max} \left(\sum_i \alpha_i \ln \pi_i(q_i; q_{-i}, c_i, d_i) + \sum_j \beta_j \ln (CS_j(P(q), a_j, b_j) - z_j) \right) \quad (3)$$

The function that is maximized in (3) will sometimes be denoted $W(q; \Theta)$. It is relatively straightforward to show that W is strictly concave in q, and therefore that (3) is a well-behaved optimization program.

Let q^* denote the value of f, the optimal vector of quota levels. Abreu and Sen's (1991) theorem guarantees that this f can be implemented virtually. A natural question is whether this result is forceful. That is, would optimal outcome q be supported under $f(\Theta)$ itself in the absence of the virtual implementing game? The answer to this question is no: exporting members *do* have an incentive to misreport their true cost parameter vector (c_i, d_i).

In the following section I devise an example with two exporters and two importers, each of whom holds preferences of the form introduced here. I calculate the outcome that would result if the exporters were to succeed in forming a pure cartel, and I show that this result cannot be supported as an equilibrium to the two-player cartel game.

An Illustrative Example

Suppose that two exporters and two importers are members of a coffee agreement that determines export quotas for each exporting member. Exporter 1 and 2 have cost functions given by, respectively:

$$C_1(q_1) = 0.03\,q_1^2$$

and:

$$C_2(q_2) = 0.05\,q_2^2.$$

Thus, exporter 1 has lower costs than exporter 2. For both exporters, $c_i = 0$. The two importing members have identical import demands, given by:

$$Q_j = 25 - 1.25P$$

so that aggregate (world) import demand is given by $Q(P) = 50 - 2.5P$. Exporter i has profits determined by:

$$\pi_i(q_i;q_{-i},c_i,d_i) = (20 - 0.4(q_1 + q_2)) \cdot q_i - d_iq_i^2,$$

where $P(Q) = 20 - 0.4Q$ is inverse world import demand. What would be the result if this pair of exporters were successfully to form an export cartel? The answer to this question is found by calculating the pair of export levels that maximizes the sum of profits across exporters. Letting $q^c = (q_1^c,q_2^c)$ denote the optimal cartel solution, the two-member cartel would export:

$$q^c = \arg \max_{q_1,q_2} \left((20 - 0.4(q_1 + q_2)) \cdot (q_1 + q_2) - \delta_1q_1^2 - \delta_2q_2^2 \right)$$

Figure 17.1 depicts the solution to this problem, which is $q^c = (14.93,8.96)$. In the figure it is clear that q^c is on the contract curve for the cartel: isoprofit curves for the two exporters are tangent at q^c. It is also clear that this cartel faces the difficulty common to quantity-setting cartels. If a member assumes that its partner will adhere to the cartel solution then each has an incentive to defect from the Agreement and overproduce.

FIGURE 17.1 Comparison of Four Quota Outcomes for Two-Exporter Example

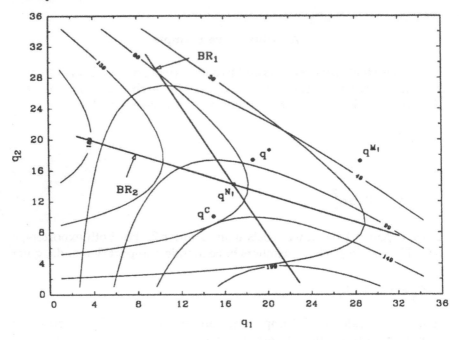

TABLE 17.1 Comparison of Four Quota Outcomes -- Illustrative Example

	Cartel Outcome	Nash Outcome	ICA Outcome	Market Outcome
q_1	14.93	16.29	18.57	28.57
q_2	8.96	14.98	17.15	17.14
$\Sigma_i\, q_i$	23.89	31.27	35.72	45.71
$P(Q)$	10.45	7.49	5.71	1.71
$\Sigma_i\, \pi_i$	238.81	215.09	178.98	88.16
$\Sigma_j CS_j$	114.06	195.57	255.19	417.96
$SW(q)$	176.43	205.33	217.09	253.06

In the figure the curves labelled $BR_i(q_{-i})$ trace out the best response correspondence for member i to export level by member -i. The intersection of the two BR_i curves, labelled q^N in Figure 17.1, illustrates the Nash equilibrium export pair. For each member, q_i^N maximizes profits *taking* q_{-i}^N *as given*. Absent some sort of enforcement mechanism, perhaps dependent upon a threat scheme, the cartel outcome is unstable.

Now, one can determine, using $f(\Theta)$ from the previous section, what would happen if the quota vector is selected by a group of members including both exporters and importers. Suppose that $\alpha_i = \beta_j = 1/2$. The solution to the resulting objective function, obtained by inserting the functional forms and the weights into (3), is denoted q^* in Figure 17.1. The figures in the final column are for the competitive market outcome, which is found by setting aggregate supply and aggregate demand equal. This outcome is efficient, maximizing the sum of consumer and producer welfare. Table 17.1 summarizes the results that obtain under the four solutions: q^c, q^N, q^* and q^M. The values in the table correspond directly to Figure 17.1. In the table, the figures for SW(q) are the sum of profits and consumer surplus, weighted by the α_i and β_j but not logged.

This simple example yields two insights that seem to shed light upon the actual workings of the ICA. The first insight is that the optimal quota vector for a pure cartel is unstable. Members have a strong incentive to defect, exporting more than their allotted amount. The second insight bears upon the following question: Why would a group of exporters join forces with their buyers--a group of importers--in an agreement like the ICA? Who gains by such an agreement? If one views the cartel outcome (q^c) and the market outcome (q^M) as two extremes, each within the realm of the possible, then it is easy to see that both groups gain. Exporters gain because they avoid the market outcome. Importers gain by ensuring that the cartel outcome is not achieved. Their influence over the quota allocations is enough to guarantee this.

Rationalizing the Behavior of the ICA

I now turn to an examination of the ICA's quota decisions during one of the last years before its collapse. The objective of this section is to *rationalize* the quota allocations that were observed during the last years of the Agreement's operation. That is, my aim is to produce a specific criterion function f, a social welfare function, such that if this function were guiding decision-making in the ICA, then the resulting optimal quota allocation would be the allocation that was actually observed. If it can be shown that f does indeed yield the behavior that occurred, then the search can be judged successful.

Given the available information on market activity and the complexity of the ICA, it is impossible to model all of the member countries explicitly. Voting rights were held by 40 exporting countries and 25 importing countries in 1985. The Agreement categorizes coffee four ways, and exporters are likewise placed in one of four categories according to the primary variety of their exports.[7] Coffee prices vary across the categories, as well as seasonally and spatially. A composite price for coffee in New York is used to monitor the world market, and that price is also employed here.

Some simplifying assumptions have been adopted to make the calculations manageable. I aggregate importing countries into three groups according to geographical location (Europe, North America, and Asia and the Pacific), and I aggregate exporting countries into four groups according to coffee variety. Thus, $n_1 = 4$ and $n_2 = 3$. I use the New York price as a market-wide price, abstracting away from price variability across varieties.

Summary of the Data

The data are for the 1984/85 crop year, during which the quotas were actually in effect. (In the following year, because of unusually high prices the quotas were suspended under the rules of the Agreement.) The following information is known from published sources: quotas held by exporters; quantities imported by members from exporting members; the voting weights as set out in the Agreement; the coffee price; and estimates of supply (demand) elasticities for exporting (importing) groups.

Part (a) of Table 17.2 summarizes the information known for exporters. Actual quota figures for each group, the q_i^*, constitute the decision made for that year by the ICA's members. Exporter vote figures are aggregated by group. The coffee price, $P = 172.04$, is the 1984 annual composite New York price, averaged from October 1984 to September 1985. Supply elasticities, denoted η_i, are taken from Akiyama and Varangis (1990). Two sets of supply elasticities are used: the short-run and the medium-run numbers presented by Akiyama and Varangis (1990). These figures give the responsiveness of supply to price changes in two years and in five years, respectively. The numbers used here are weighted averages of the country elasticities.[8] Knowing the supply elasticity figures and the functional form for marginal cost, the c_i may be expressed in terms of d_i for each exporter, thereby reducing the number of free parameters for each exporter to one. These are given by:

$$c_i = 2d_iq_i(\eta_i - 1) \tag{4}$$

Knowing q_i and η_i, if the d_i can be found then one knows fully the marginal cost function and, hence, the cost and profit functions for i.

The corresponding set of information for importers is presented in part (b) of Table 17.2. Import quantities are imports by members from exporting members. Because the reported data do not agree exactly in any one year (due to changes in inventory that are built into the import figures), these figures are approximate. They have been scaled proportionally to ensure that total exports and total imports agree. The resulting error should be negligible. Importer vote figures are sums for countries belonging to the respective groups, out of a total of 1000 for importers.

Akiyama and Varangis (1990) provide the data used to calculate group import demand elasticities, denoted ϵ_j. They are weighted averages of member countries. Knowledge of demand elasticities, together with a price-quantity pair (hypothesized to lie on the group's demand curve) for each group allows the direct calculation of the demand function parameters a_j and b_j. The derivation follows the lines of that for c_i above; the difference is that in the case of the importers a point on the demand curve is known. This permits one to specify the demand function exactly.[9]

Everything about the criterion function (3) is now known, except the d_i. Calculating them is the key numerical task of the paper.[10] The d_i can be derived from the criterion function f. A great deal is already known about f. Most importantly, it is a weighted sum of logged profits (for exporters) and logged consumer's surpluses less the importer threat points (for importers). The actual price P is known, the aggregate demand function is known, and the quota vector q^* that it is to yield at a solution is known. If the correct d_i are found, they must be such that the resulting f achieves a maximum at the actual q^*.

The criterion function from equation (3) is, once again:

$$W(q;\Theta) = \Sigma_i \alpha_i \ln(\pi_i(q;c_i,d_i)) + \Sigma_j \beta_j \ln(CS_j(q;a_j,b_j) - z_j)$$

Under the assumption that P is less than the smallest a_j (an assumption that is violated only if one or more importers buy a quantity of zero), the relevant range of demand is:

$$P(q) = \left(\frac{1}{\Sigma_j (1/b_j)} \right) \cdot \left(\Sigma_j \frac{a_j}{b_j} - \Sigma_i q_i \right).$$

Preferences for exporters and importers are as given above. $W(q;\Theta)$ is found by inserting π_i and CS_j into the appropriate places. At an

TABLE 17.2 Export and Import Data by Group

	(a) Exporting Groups (Quantity figures in millions of 60kg bags)			
	Columbian Milds[a] (i=1)	Other Milds[b] (i=2)	Brazilian Arabicas[c] (i=3)	Robustas[d] (i=4)
η_i (Short Run)	0.1441	0.1066	0.0324	0.2818
η_i (Medium Run)	0.4003	0.1490	0.1040	0.3815
Actual q_i [e]	11.33	14.31	18.86	15.11
Votes[e]	183	264	277	276

	(b) Importing Groups (Quantity figures in millions of 60kg bags)		
	Asia and Europe[f] (j=1)	North Pacific[g] (j=2)	America[h] (j=3)
ε_j	-0.2157	-0.3135	-0.4305
a_j	969.54	720.79	782.52
b_j	-2.397×10^{-5}	-10.930×10^{-5}	-2.864×10^{-5}
Q_j [i]	33.273	5.02	21.32
Votes	599	101	300

		Coffee Price	P=172.04/bag

Notes:
a. Colombian Milds—(Votes in parenthesis): Colombia (142); Kenya (25); Tanzania (16).
b. Other Milds: Bolivia (6); Burundi (6); Costa Rica (23); Cuba (6); Dominican Republic (0); Ecuador (21); El-Salvador~(40); Guatemala (34); Honduras (18); India (16); Jamaica (4); Malawi (4); Mexico (35); Nicaragua (14); Panama~(4); Papua New Guinea (14); Paraguay (6); Peru (0); Venezuela (4); Zimbabwe (4).
c. Brazilian and Other Arabicas: Brazil (252); Ethiopia (25).
d.   Robustas: Angola (9); Ghana (4); Indonesia (45); Liberia (6); OAMCAF (includes: Cameroon (23), Central African Republic (4), Ivory Coast (56), Madagascar (12), Togo (5)); Philippines (11); Rwanda (11); Sierra Leone (8); Sri Lanka (4); Thailand (6); Uganda (42); Zaire (22); Zambia (4).
e. Source: USDA (January, 1985).
f. Europe: Austria (20); Belgium/Luxembourg (32); Cyprus (6); Denmark (20); Finland (20); France (87); FR Germany (134); Greece (11); Ireland (6); Italy (62); The Netherlands (43); Norway (15); Portugal (9); Spain (31); Sweden (29); Switzerland (20); United Kingdom (42); Yugoslavia (12).
g. Asia and the Pacific: Australia (15); Fiji (5); Japan (58); New Zealand (7); Singapore (16).
h. North America: Canada (32); United States (268).
i. Source: USDA (February, 1987).

optimum, the first order necessary conditions for a solution to the ICA's problem require that the four first partial derivatives of W be zero. These derivatives are:

$$\frac{\partial W}{\partial q_i} = \frac{\alpha_i}{\pi_i}\left[(P - c_i)\sum_{j=1}^{3}(1/b_j) - q_i(1 + 2d_i\sum_{j=1}^{3}(1/b_j))\right]$$
$$- \sum_{k=1}^{3}\frac{\alpha_k q_k}{\pi_k} + \sum_{j=1}^{3}\frac{\beta_j Q_j}{(CS_j - z_j)} = 0,$$

(5)

for i = 1,. . . 4. The only unknowns in equations (5) are the d_i. Thus, these equations constitute a four-equation system in four unknowns, the d_i. The solution to this system yields the remaining unknowns in Θ, the vector of preference parameters. The empirical results of the study, involving the solution to equations (5), are now presented.

Empirical Results

The results of solving equations (5) numerically are presented in Table 17.3, where the d_i^* denote the solution, and where the c_i^* are calculated from equation (4). Part (a) of Table 17.3 contains the results using the short-run export elasticities. Part (b) of Table 17.3 contains the results using medium-run export elasticities. The marginal cost values in the table are calculated from the derivatives of each exporter's cost function evaluated at q_i^*. Because this is not a competitive outcome, marginal costs for exporters need not equal each other or the actual coffee price. They do not. Recall that the actual coffee price is P = 172.04.

In the short run, the Brazilian arabicas group has the lowest export elasticity, and also has the flattest marginal cost. Brazil enjoys a low marginal cost at its output level. The marginal cost figures are to be compared to the coffee price of P = 172.04. Marginal cost is relatively high for the Robustas group, whose export elasticity is also quite high. In the medium run, the Brazilian arabicas group once again has the lowest marginal cost. The Robustas groups once again has the highest marginal cost. All of the marginal cost figures are still well below the coffee price, indicating that the exporting countries earned a positive profit at the quota export levels.

The preference orderings given by the consumers surplus measures for importers and by the profit functions (incorporating the calculated d_i^* and c_i^*) by construction yield a *rationalizing* f. All of the required information

TABLE 17.3 Cost Function Parameters: Empirical Results

	(a) Short Run Elasticities			
	Colombian Milds ($i=1$)	Other Milds ($i=2$)	Brazilian Arabicas ($i=3$)	Robustas ($i=4$)
d_i^*	6.2327×10^{-6}	7.0747×10^{-6}	2.9248×10^{-6}	5.4612×10^{-6}
$c_i^* = 2d_i\, q_i(\eta_i - 1)$	-120.850	-180.922	-106.730	-118.568
MC_i^*	20.350	21.592	3.573	46.523

	(b) Medium Run Elasticities			
	Colombian Milds ($i=1$)	Other Milds ($i=2$)	Brazilian Arabicas ($i=3$)	Robustas ($i=4$)
d_i^*	10.148×10^{-6}	13.992×10^{-6}	8.8360×10^{-6}	8.5653×10^{-6}
$c_i^* = 2d_i\, q_i(\eta_i - 1)$	-137.874	-340.875	-298.586	-160.106
MC_i^*	92.031	59.661	34.651	98.736

(the Θ vector of the c_i, d_i, a_j, and b_j representing preferences) is now available. Verifying that the q_i are rationalized by the cost and demand functions derived here involves locating the maximum to $W(q;\Theta^*)$, a calculation that was carried out and shown to provide the anticipated result. On this count the exercise has been successful.

It is now possible to use these results to investigate another suite of questions. Specifically, the set of comparisons that was presented in the example above may now be made for the ICA. All of the required elements are in place. Price, quantity, and welfare comparisons between the ICA outcome and the Nash, cartel, and competitive outcomes may be conducted. All of the relevant calculations were carried out for both the short-run and the medium-run export elasticities, and are presented in Table 17.4. As before, SW(q) denotes the weighted sum of unlogged profits and consumer surpluses.

The cartel outcome q^c is the solution to the problem:

$$\max_q \sum_{i=1}^{4} \pi_i(q_i; q_{-i}, c_i, d_i).$$

TABLE 17.4 Comparison of four quota outcomes--International Coffee Agreement

	(a) Short Run Elasticities			
	Cartel Outcome	Nash Outcome	ICA Outcome	Market Outcome
q_1 [a]	7.16	12.92	11.33	13.36
q_2 [a]	10.56	14.40	14.31	16.01
q_3 [a]	12.85	16.99	18.86	26.05
q_4 [a]	7.97	13.70	15.11	15.03
$\Sigma_i\, q_i$ [a]	35.54	58.01	59.61	70.45
$P(Q)$ [b]	417.67	190.72	172.04	45.65
$\Sigma_i\, \pi_i$ [c]	19,250	14,291	13,481	6,145
$\Sigma_j\, CS\_j$ [c]	9,098	20,053	21,152	29,372
$SW(q)$ [c]	9,461	13,165	13,486	15,760

	(b) Medium Run Elasticities			
	Cartel Outcome	Nash Outcome	ICA Outcome	Market Outcome
q_1 [a]	5.16	11.09	11.33	11.82
q_2 [a]	11.00	14.06	14.31	15.83
q_3 [a]	15.02	17.56	18.86	22.67
q_4 [a]	7.42	13.08	15.11	15.31
$\Sigma_i\, q_i$ [a]	38.60	55.80	59.61	62.62
$P(Q)$ [b]	416.91	216.49	172.04	102.01
$\Sigma_i\, \pi_i$ [c]	21,800	17,535	13,268	7,770
$\Sigma_j\, CS\_j$ [c]	9,127	18,587	21,152	29,372
$SW(q)$ [c]	10,250	13,409	14,058	15,291

Notes:

[a] Million bags.
[b] $US.
[c] Million $US.

The Nash outcome q^N is the joint solution to the four problems (one for each exporter):

$$\max_{q_i} \pi_i(q_i; q_{-i}, c_i, d_i)$$

where each exporter takes the exports of others as fixed quantities, treating them parametrically. The third comparison is between the ICA outcome and the market outcome, denoted q^M. The market outcome is obtained by solving for the joint solution to aggregate supply and demand.

Part (a) of Table 17.4 presents the four sets of results from the short-run elasticity case. At q^c the coffee price would be $417.67 per bag—more than twice the actual observed price. Exporters, naturally, would gain handsomely if this result could be achieved, but as in the example it is not self-sustaining. Each exporter would have a strong incentive unilaterally to defect away from the cartel solution.

Consumer surplus for importers is quite low at the cartel solution, and so is aggregate (weighted) social welfare. If each exporter takes the export *quantities* of others as given, the Nash solution obtains, and the price would be $190.72 per bag. Total exports and social welfare are only slightly lower than at the ICA outcome, but exporters are better off and importers worse off.

This market equilibrium is achieved for $\Sigma_i\, q_i$ = 70.45 million bags, and the price per bag is $45.65. Exporters each take this *price* as given and maximize profits by choosing the export level q_i^M at which MC_i = P. Of the four cases, this outcome yields the lowest profits to exporters, the highest surplus to importers, and the greatest overall social welfare.

Part (b) of Table 17.4 presents the corresponding set of results for the medium-run export elasticities case. The relative values for export quantities, prices, profits and consumer surpluses, and social welfare remain the same. There are some interesting differences, however. The most striking difference is that the competitive export quantities are lower and the competitive price higher than in the short-run case. The price of $102 per bag is still considerably below the actual price of $172.04, but it is more than double the competitive price in the short run.

Even though the results were not obtained using a statistical procedure, and thus cannot be used to make forecasts, still there is a sense in which the market outcomes in Table 17.4 constitute a set of projections about what would happen if there were no ICA. An interesting question comes to mind regarding the degree to which one can trust these numbers. Fortunately, the 1989 demise of the Agreement provided an opportunity to test whether they bear any resemblance to reality. In the four years since the Agreement ceased to function, the market has had time to move toward a competitive situation.

The World Coffee Market in 1990/91 and 1991/92

The December 1992 issue of the USDA-FAS publication "World Coffee Situation" (USDA 1992, p. 9) reports exports by member exporting countries. As a final exercise I compare the market outcome projected by the model to actual figures from the 1990/91 and 1991/92 market years. These are the most recent years for which the required information is available. Though the comparison made in this section is a useful one, an important *caveat* should be mentioned. The model of this paper is static; the world coffee market is not. The model can say only what would happen at a competitive equilibrium if the supply and demand functions that I have derived are the correct ones, and so the exercise is perhaps best thought of as tracing out the direction that the market might take as it moves forward in time. Even if the "projected" outcomes are correct the model cannot say when they should obtain.

Table 17.5 contains the numbers that permit comparison between the predicted and actual (1990/91 and 1991/92) quantity and price figures. The first two columns in the table present the projections from the short- and medium-run calculations, respectively. Given that the short-run elasticities contain information on the market response that might occur in two years or so, they provide the most reasonable comparison to the 1991/92 figures. The comparison is by no means perfect, although the aggregate export quantities are not far wrong. The model projects total exports from member exporters to member importers of 70.45 million bags, while actual exports were 66.50 million bags. This is a discrepancy of 5.61 percent. The model projects that the price should be $45.65 per bag, while the actual 1991/92 price was $57.90 per bag. This is a discrepancy of 21.16 percent. Brazil's production increased by less than expected, while Colombian Milds increased by more than expected. The discrepancy for Brazil might be due to the domestic credit program that has encouraged farmers to hold stocks in the hope that prices would rise. Colombia continued its generous subsidy program, shielding its growers from world prices. Without the quota restriction and in the interest of maintaining foreign earnings, Columbian coffee exports have been quite high in the last two years. Exports of robustas and of Other Milds moved upward from the 1989 demise of the ICA to 1991/92 in amounts very similar to those of the projections.

It is not so easy to interpret the results of the medium-run elasticity case, for too little time has passed to provide a fair test of those figures.

TABLE 17.5   A Comparison Between Actual and Projected Market
Outcomes

	Market Outcome (SR Elast)	Market Outcome (MR Elast)	1990/91 Actual[a]	1991/92 Actual[a]
Columbian Milds	13.36	11.82	12.724	15.045
Other Milds	16.01	15.82	16.720	17.209
Brazilian/Arabic as	26.05	22.67	17.384	18.568
Robustas	15.03	15.31	16.667[b]	15.156
Member Totals	70.45	63.63	62.505[c]	66.498
Price Per Bag ($US)[d]	45.65	102.01	74.20	57.90

Notes:
[a] *Source*: ICO.
[b] Includes Vietnam.
[c] Equals exports by ICO exporting members to all destinations, multiplied by the fraction of all imports accounted for by member importers.
[d] 1990/91 and 1991/92 figures are average for October–September, deflated by the GAP deflator.

The medium-run elasticities show the response that would obtain in five years, and so their accuracy must await the passage of another two years or so. It is somewhat worrying that the projected price is *higher* in the medium-run case than in the short-run case. This may be due to the fact that though supply curves are flatter the marginal cost of growing coffee also increases over time as growers are required to plant new trees and otherwise elevate their capital investments.

Conclusions

How does trade occur among nations who explicitly set out to act in concert? In seeking to answer this question one can hardly avoid the attendant problem of the nature of their decision-making procedure. It would seem there is much to be gained by modeling such a collective body as an optimizing entity, but there is also some danger in doing so. Social choice theory tells us that a group of deciders may have no well-defined will or objective criterion. In this paper I have constructed a criterion for decision-making by the International Coffee Agreement. The objective criterion I present, if true, would have produced the decisions that were actually observed for the 1984/85 crop year. I offer it up as a candidate objective for the ICA.

The objective function was itself created from the data, using price and quantity information and demand and supply elasticities for member countries. Clearly, further work is required in order to shed light upon the dynamic nature of the coffee market and its various details. Still, the leading contribution of the paper is to show that it is possible to devise a collective optimizing program for a group of countries without doing violence to our mathematical or logical sensibilities.

Notes

1. This chapter draws upon Coggins 1995,where the model is developed in greater detail.
2. In the case of the ICA this is not quite true. Not all trading countries are members, and member exporters sometimes evade their quotas by selling to a non-member, who in turn exports to a member importer. Coffee that travels this circuit is called "tourist coffee" by the trade.
3. This section follows Abreu and Sen (1991).
4. This index can be quite general. It might correspond to a parameter in a function describing preferences, or, in the case of exporters, parameters in the member's cost function. In the latter case, with demand given the member's preferences correspond to its profit function defined on Q.
5. Strictly speaking, in Abreu and Sen (1991) the value of f is $\mathfrak{L}$, a lottery over Q. I will leave this technical distinction aside and call Q the range space of f.
6. Throughout the remainder of the paper q_i denotes a quota amount (or, equivalently, exports) for exporter i and Q_j denotes imports by importer j. By definition it will always be true that $\Sigma_i q_i = \Sigma_j Q_j$. The symbol Q denotes the sum.
7. This construction is lifted directly from the ICA, under which importing members are given 1000 votes and exporting members are given 1000 votes.
8. The four are "Colombian milds," produced in Colombia predominantly; "Other milds," produced in most of the rest of Latin America; "Brazilian and other arabicas," produced in Brazil and Ethiopia; and "Robustas," produced mostly in Africa. See Wrigley (1988).
9. Akiyama and Varangis (1990) do not estimate supply elasticities for every country that is a member of the Agreement. My figures for each group are weighted averages of the group members whose country elasticities are given in Akiyama and Varangis.
10. This point is important, and perhaps bears some elaboration. It is true that exporters receive the same price for their product that importers pay. However,

while this price lies on an importer's demand curve (at the corresponding quantity) it needn't lie on an exporter's supply curve.

11. Note that though the groups' preferences (in the case of an exporter, a profit function) are what we are after, the calculation to follow yields a cost function. However, information on coffee demand, which is also a part of each profit function, is already known fully.

References

Abreu, D. and A. Sen. 1991. "Virtual Implementation in Nash Equilibrium." *Econometrica* 59:997-1021.

Akiyama, T. and P. N. Varangis. 1990. "The Impact of the International Coffee Agreement on Producing Countries." *World Bank Economic Review* 4:157-173.

Arrow, K. 1962. *Social Choice and Individual Values* New Haven, CT: Yale University Press.

Baldwin, R. E. and R. N. Clarke. 1987. "Game-Modeling Multilateral Trade Negotiations." *Journal of Policy Modeling* 9:257-284.

Coggins, J.S. 1995. "Rationalizing the International Coffee Agreement Virtually." *Review of Industrial Organization*. Forthcoming.

Donsimoni, M. P., N.S. Economides, and H.M. Polemarchakis. 1986. "Stable Cartels." *International Economic Review* 27:317-327.

Fisher, B. S. 1972. *The International Coffee Agreement: A Study in Coffee Diplomacy* New York, NY: Praeger Publishers.

Gibbard, A. 1973. "Manipulation of Voting Schemes: A General Result." *Econometrica* 41:587-601.

Gilbert, C.L. 1987. "International Commodity Agreements: Design and Performance." *World Development* 15:591-616.

Gordon-Ashworth, F. 1984. *International Commodity Control: A Contemporary History and Appraisal.* New York, NY: St. Martin's Press.

Green, E.J. and R. H. Porter. 1984. "Noncooperative Collusion Under Imperfect Price Information." *Econometrica* 52:87-100.

Harrison, G. W. and E.E. Ruström. 1991. "Trade Wars, Trade Negotiations, and Applied Game Theory." *Economic Journal* 101:420-435.

Karp, L. S. and J. M. Perloff. 1993. "A Dynamic Model of Oligopoly in the Coffee Export Market." *American Journal of Agricultural Economics* 75:448-457.

Marshall, C.F. 1983. *The World Coffee Trade: A Guide to the Production, Trading and Consumption of Coffee.* Cambridge, MA: Woodhead-Faulkner.

Maskin, E. 1977. "Nash Equilibrium and Welfare Optimality," mimeo, MIT, Cambridge, MA.

Matsushima, H. 1988. "A New Approach to the Implementation Problem." *Journal of Economic Theory* 45:128-144.

Moore, J. 1992. "Implementation, Contracts, and Renegotiation in Environments with Complete Information," in J.J. Laffont, ed., *Advances in Economic Theory: Invited Papers for th Sixth World Congress of the Econometric Society, VI* Cambridge: Cambridge Univesity Press.

Osborne, D.K. 1976. "Cartel Problems." *American Economic Review* 66:835-844.

Pieterse, M.T.A. and H.J. Silvis. 1988. *The World Coffee Market and the International Coffee Agreement*. Wageningen, Netherlands: Wageningen.

Porter, R. H. 1983. "Optimal Cartel Trigger Price Strategies." *Journal of Economic Theory* 29:313-338.

Satterthwaite, M.A. 1975. "Strategy-Proofness and Arrow's Conditions: Existence and Correspondence Theorems for Voting Procedures and Social Welfare Functions." *Journal of Economic Theory* 10:187-217.

Stigler, G.J. 1964. "A Theory of Oligopoly." *Journal of Political Economy* 72:44-61.

United States Department of Agriculture. 1985. "World Coffee Situation." Circular Series FCOF 1-85.

United States Department of Agriculture. 1987. "World Coffee Situation." Circular Series FCOF 1-87.

United States Department of Agriculture. 1992. "World Coffee Situation." Circular Series FCOF 3-92.

Wrigley, G. 1988. *Coffee*. New York, NY: Longman Scientific and Technical.

18

Conclusion

Philip C. Abbott and Ian M. Sheldon

A number of important changes have been on-going in international agricultural markets, and especially for trade in processed food products. Over the last fifteen years, high-value agricultural trade has been growing substantially more rapidly than trade in bulk commodities. The value of U.S. exports of high-value agricultural products now exceeds that of commodity exports, a situation reached much earlier in the European Union (MacDonald and Lee 1994). This trend toward greater trade in high value products must be kept in perspective, as Handy and Henderson (1994) note, domestic sales of foreign owned subsidiaries of U.S. multinationals are roughly four times the magnitude of exports for processed food products. Thus, the tendency to invest abroad is much stronger than to service foreign markets by exporting. This alternative to trade is an important and growing phenomenon, and may be a complementary rather than competitive firm strategy.

These trends in international trade and foreign investment by the U.S. food industry mirror the empirical crises that have led to new developments in the theory of international trade, and similarly in the explanations underlying foreign direct investment. Trade in all goods is becoming increasingly intra-industry. That is, similar products are traded between similar countries, contrary to the predictions of Heckscher-Ohlin trade models. Trade is often conducted between subsidiaries of a multinational firm (intra-firm trade) rather than as "arm's length" transactions. Foreign direct investment (FDI) is also moving in a similar direction (Ethier 1994). Early trade theories depended on differing resource endowments as the principal explanation of observed trade patterns. New theories have introduced the concepts of imperfect competition, product differentiation, and economies of scale. Similar new developments in underlying theory have also occurred in the industrial organization area, where the relevance of these same factors to market behavior has also been recognized (Carlton and Perloff 1994). Indeed, much of what is done in the new trade theory seeks to incorporate developments from the new industrial organization theory into trade models (Krugman 1989).

The methods employed in those models are, of necessity, dependent upon the institutional structures of the markets investigated. Furthermore, there has been a continuing evolution of methods to address those problems in the theoretical literature, with a number of alternative proposals put forward. As a consequence, the empirical investigations in this book, which draw upon these bodies of literature, reflect the diversity of approach found there. Some of the studies in this book look very much like applications of either the new trade theory or new industrial organization theory. Others draw upon some of the earlier methods (such as the structure-conduct-performance – SCP -- tradition in industrial organization (Bain 1951)) to understand important factors driving observed trends in international markets for agricultural commodities and processed food products (e.g., Pagoulatos and Sorenson 1976).

Several chapters in this book focused on providing a deeper understanding of events in these markets, whereas others focus on the application of a particular new method at the frontier in these areas to a specific market situation. The unifying theme of the chapters is that the relevance of imperfect competition and the factors leading to imperfect competition, including economies of scale and product differentiation, are considered as critical elements of the models and methods employed.

The first set of chapters in this book examined the extent of imperfect competition in international food markets and the factors underlying the existence of those market forces. The book then moves to considering some of the influences of policy under imperfect competition, and how we might go about measuring policy impacts in that economic environment. FDI and the role of multi-national firms, including intra-firm trade, are examined next. The concluding chapters look at these influences in specific markets, using some of the most sophisticated and analytical methods now available.

In chapter 2, Field and Pagoulatos examine relationships between foreign trade, domestic market structure, business cycles, and profit margins for U.S. food manufacturing industries. They develop a model of oligopoly markets with international trade and empirically estimate that model for 43 SIC four digit U.S. food manufacturing industries over the sample period 1972 to 1987. Their model emphasizes the role of the price elasticity of demand in assessing consequences of imperfect competition in the food industry. Key findings are support for the negative influence of the price elasticity of demand on profit margins and the important role of seller concentration in exerting a positive effect on price-cost margins. They also find a positive relationship between import share and profit margins, while exports exert a negative effect on U.S. food industry profit margins. Imports work both through their effect on competition and on concentration, since intra-firm trade is important to these firms. Exports lead to more domestic firms, greater competition, and to reduced margins. The counter cyclical nature of food industry profit margins suggests an advantage of these firms during recessions which merits further study.

Patterson, Reca, and Abbott investigate price discrimination in international markets by U.S. high value food product exporters in chapter 3. They use a variant on the pricing-to-market (PTM) model, introduced by Krugman (1987), and implemented by Knetter (1989), to examine the extent of PTM behavior for the U.S. chicken and beef industries. They also develop a model following the SCP tradition of the industrial organization literature, based on a model developed by Lyons (1981), which relates export price markups to concentration along trade routes. They also modify the PTM model to account for the factors determining price discrimination in Lyons' (1981) model. They show the relationship between these two models -- the measure of market power found in the PTM model is related to industry concentration, a measure found to be important in the SCP literature. Their empirical results provide some evidence on discriminatory pricing by destination in the chicken export sector. In their work on U.S. beef exports, using more disaggregated data, they demonstrate that product heterogeneity may be a more important reason for price discrimination than imperfectly competitive behavior by firms. This chapter also notes that exchange rates are related to more than just pricing relationships in net import demand models, so this may confound the relationships examined using the traditional PTM approach.

In the following chapter, Pick and Park present an econometric method for improving upon Knetter's (1989) approach to testing for pricing-to-market (PTM) in U.S. agricultural exports markets. They use a symmetry condition applied to adjustments in the optimal price-cost markup by destination to increase estimation efficiency. Their empirical work on export pricing decisions for U.S. wheat exports supports the existence of PTM behavior in 6 of 8 destinations when the symmetry condition is imposed. A model which relaxes that restriction produces inferior econometric results, demonstrating the advantage of this approach. Their results also confirm the importance of the Export Enhancement Program (EEP) in determining destination specific wheat prices.

Lopez and Pagoulatos then examine political economy factors underlying observed trade barriers in the food processing industry. They observe that these barriers are more restrictive than those found in most manufacturing industries. The potential to erect such barriers causes firms to compete in the political arena to influence the level of such trade barriers, instead of concentrating their efforts on normal market competition. They also consider the relationship between market structure characteristics, including the level of concentration, and factors determining comparative disadvantage, on the magnitude of trade barriers across U.S. food processing industries. Their empirical results for forty-four food and tobacco manufacturing industries demonstrate the importance of including direct political variables, especially political action committee (PAC) contributions, into models explaining observed protection. They also show that trade protection decreases with seller concentration according to a non-linear relationship. The relationships between protection and employment or comparative disadvantage are weaker, and those variable alone do not explain well observed trade barrier patterns in U.S. food processing industries.

In chapter 6 Hertel develops a new measure of competitiveness which is subsequently used to examine factors behind the competitiveness of U.S. food exports to Japan. His index is based on simulations using a global computable general equilibrium model constructed from the SALTER data base (Jomini, *et al.* 1991), which captures U.S., Japanese, and competing exporting firm responses. Important factors examined there include current import barriers, the unit costs of both food and non-food production, and alternative assumptions about market structure. The basis for his alternative measure is consumers' unit expenditure, which is made an explicit function of consumer heterogeneity, reflecting inherent biases for Japanese products, the number of varieties available in Japan, as well as relative prices. Theoretical details behind this measure are developed and an empirical application to U.S. food exports to Japan is provided. The competitiveness of U.S. food exports are found to move closely with tariffs when entry is permitted in markets. That relationship is attenuated somewhat when entry is prohibited, demonstrating the procompetitive effects of entry/exit by firms. Technical change in the U.S. is also considered, as is the situation where global improvement in food manufacturing efficiency occurs. Incomplete price transmission, as a result of international transportation costs, leads to an erosion of U.S. competitiveness when there is global technical change. This general equilibrium approach also highlights the importance of technical change in the Japanese non-food sector in determining competitive advantage vis-a-vis U.S. food exports, which has been more rapid than for food manufacturing, leading to a deterioration of Japanese domestic food manufacturing competitiveness.

In chapter 7 van Duren and de Paz investigate the impact of the Canada-U.S. Trade Agreement (CUSTA) on Canada's tomato processing industry. They observe that this industry continues to struggle for survival, given changes in tariffs, technical regulations, and contract bargaining behavior that have followed CUSTA. They used a multi-region, multi-product spatial equilibrium model of the North American tomato industry to investigate the competitiveness of this sector in Canada. They find that reduction of tariffs due to CUSTA greatly reduced the protection to this Canadian industry, and the cumulative effect of technical regulations imposed on the industry have also substantially reduced competitiveness. They did not find that the ongoing harmonization of technical regulations was as likely to be as serious a problem as the tariff reductions of CUSTA, however. These results are strongly conditioned by the bargaining behavior of industry participants, which influences both pricing mechanisms and productivity.

The relationship between trade liberalization and productivity growth is considered by Kalaitzandonakes and Bredahl in chapter 8. They find, using empirical evidence from agricultural industries which have been recently liberalized, that economic reform need not necessarily cause a shrinkage in a previously protected sector. Reduction in protection can lead to increased production when it is accompanied by important domestic reforms, when the domestic policies more negatively impact the protected sector, and when the removal of protection brings renewed investments by better managed operations. They develop a theoretical model to demonstrate that protectionism can induce inefficiency through effects on scale economies, managerial effort, and investment decisions. That model is implemented along the lines proposed by Mundlak (1988) to capture endogenous technical change. Their empirical evidence, taken from the New Zealand sheep/beef industry, highlights this unexpected result, and emphasizes the inefficiencies introduced by domestic regulation and the lack of competition internally. When protection was lifted and imports permitted, and competition intensified, productivity could grow in some domestic industries. Evidence on investment patterns and productivity corroborated expectations from their theoretical model for a high income industry.

In chapter 9 Hirschberg and Dayton observe that a significant proportion of world trade in processed food is increasingly composed of intra-industry trade. They provide a detailed statistical analysis of the factors explaining observed patterns of intra-industry trade in processed food products. They use a clustering technique to examine the degree to which this trade is a function of inherent technology and market structure factors, specifically examines the Helpman-Krugman (1985) hypothesis that intra-industry trade is more likely between similarly endowed countries. They identify specific industries which are significantly more prone to engage in intra-industry trade as a result of increasing similarity in technology across countries. While their aggregate results also support the Helpman-Krugman (1985) hypothesis, some clusters of sectors are found for which factor endowments or environmentally based advantages better explain observed trade.

In chapter 10 Wang and Connor investigate underlying explanations of intra-firm international trade (IFT) in processed food products. The purpose of their paper is to test the relative predictive accuracy of three mutually inconsistent theories explaining IFT. These paradigms are transactions costs, industrial organization, and transfer pricing approaches. They apply their tests to data on all U.S. manufacturing industries, paying particular attention to food product trade. They find strong evidence in support of market structure models, rather than models based on internationalization theories, in explaining both intra-firm exports and imports. More consistent results in predicting IFT imports and exports are found for food industries than for general manufacturing, leading them to conclude that IFT in food products is notably different form other manufactured goods trade. Strong results are found for all market structure variables, suggesting that in the food industry less concentrated industries producing highly differentiated products engage in IFT.

In chapter 11 M. Reed and Ning investigate foreign direct investment strategies of multinational food firms as an alternative means of gaining foreign market share. Strategies followed by U.S. food firms are analyzed using two methods. Case studies of five important U.S. multinational enterprises (Sara Lee, Borden, Campbell Soup, CPC International, and Kellogg) are presented. Then regression analysis is used to examine the determinants of foreign direct investment for thirty-four food processing firms. From the five case studies, they observe that two firms with diversified sales, which included nonfood items, have fared better than the three firms specializing in food product exports. All firms had increased their foreign sales more rapidly than their U.S. sales. Their findings from the regression analysis support the contention that advertising and marketing by American firms are important determinants of success of foreign direct investment, and that foreign direct investment is a diversification and market access strategy of these firms. They also found that firms with a substantial involvement in research and development may prefer to export products rather than risk having their technology flowing to foreign countries as a consequence of direct foreign investment.

In chapter 12 Henderson, Vörös, and Hirschberg examine industrial organizational characteristics as possible determinants of the extent to which food processing firms pursue foreign direct investment versus export strategies. Their paper tests hypotheses relating the intensity of FDI by food and beverage manufacturing firms to firm dominance in its home market, to product diversity or specialization and to investment in intangible firm specific assets. Firm level data on 312 firms for FDI and 105 firms to examine export propensity were taken from *Global Company Handbook* published by the Center for International Financial Analysis and Research (CIFAR 1992). They also consider the extent to which exports and foreign direct investment are strategic behaviors which enhance profits. Results confirm finding of earlier studies and prior theoretical expectations. They find that firms dominant in their home market tend to invest more abroad and export less. Furthermore, smaller firms are more inclined to export, whereas larger firms exhibit greater FDI intensity. Investment in intangible firm specific assets as a direct cause of foreign direct investment is also demonstrated. Specialization by firms in food or beverage manufacturing enhances the propensity to export, but has no significant effect on the pattern of FDI. Product diversity, on the other hand, encourages FDI. They argue that the observed dominance of FDI relative to exports by these firms is due to the profit enhancing consequences of that strategy.

In chapter 13 A. Reed provides another empirical investigation of the determinants of FDI in the processed food industry. A model is developed to explain FDI which is based on firms exploiting comparative management skill, realizing economies of scale or size, and acquiring more precise information on consumer behavior. A Kalman filter procedure is used to update forecasts of market behavior where some data is unobservable, and bootstrapping is used to compute standard errors on more precise parameter estimates. The empirical results examine the relative importance of these three advantages, based on strategy changes in response to consumers. They determined that, from the firm's perspective, the gain in information on a market from undertaking an FDI strategy results in a smoother supply response than if the firm exclusively pursues an export strategy. Locating firms closer to target markets provides better consumer data on which to make such decisions. Economies of scale are also important to firms undertaking FDI, enabling them to become less sensitive to demand shocks.

McCorriston and Sheldon then examined the consequences of agricultural policy reform in trade models, specifically considering the consequences of different trade policy instruments -- tariffs versus quotas -- in the context of oligopolistic markets. They develop a general conjectural variations model that captures the anticipated effects of tariffs and quotas in an imperfectly competitive setting and apply that model to banana imports by Germany under the European Union's (EU) common agricultural policy. Their theoretical model shows how tariffs and quotas are non-equivalent in their impacts on firm behavior, potentially impacting conjectures held by firms. Their empirical analysis highlights the importance of who receives the tariff revenue and who benefits from the rents generated by quotas. They also argue that the negative, anti-competitive effects of protection outweigh the tariff revenue effects, cautioning that pro-competitive effects could be a dominant feature in other markets. Welfare losses in this case also depend strongly on product differentiation.

Bourgeon and Le Roux also consider the consequences of agricultural policy in the European Union in an imperfectly competitive setting. They use auction theory to consider the effects of the grain policy regime on French wheat exports, focussing on the process by which EU export refunds are awarded. They examine empirically the implications of their method, considering tender offers involving soft wheat intervention stocks. While they note that the empirical results do not yet provide clear conclusions, they are able to highlight several important features of the EU grain market. In particular, the strong relationship between European trader behavior and U.S. export prices is observed, confirming the importance of the U.S. as a price leader in the wheat market, and the use of the EEP to play that role. Their methods also employ one of the approaches at the frontiers of modeling trade in imperfect markets (auction theory), observing that more work is needed on these methods to model circumstances found in actual food markets. They note the need to develop structural models based on theory which fit more closely the particular markets to be analyzed.

In the next chapter Réquillart, Cabolis, and Giraud-Héraud focus on the welfare effects of the introduction of high fructose corn syrup (HFCS) into the U.S. sweetener market. They consider the substitutability between HFCS and sugar, and the extent to which U.S. trade policy (which has raised the price of sugar via quotas) has fostered the development of HFCS. They emphasize that induced technical change has led to improvement in the properties of HFCS -- citing the importance of increasing the sweetness of HFCS which made this a more acceptable substitute for sugar. They employ a differentiated product model in the context of imperfect competition to examine the derived demand for sweeteners as an intermediate input, deriving the welfare effects from U.S. sugar policy with their models. Their results yield smaller impacts on welfare resulting from U.S. sugar protection than have been found in previous studies.

In the final chapter, Coggins considers outcomes in the world coffee market as a consequence of the International Coffee Agreement (ICA), explained in an imperfectly competitive setting. The problem he poses is to determine, given information on the preferences of cartel members, the levels of export quotas allocated to individual members of the coffee agreement. He models this social choice problem using a "virtual implementation" of a Nash equilibrium among cartel markets. (Abreu and Sen 1991). His problem goes beyond the standard analysis of a cartel of producing countries by including in the decision process agents who are also importers. This reflects the actual structure of the ICA. His theoretical model provides insight into the likely outcomes under an inherently unstable cartel. He then uses his model to examine empirically coffee market outcomes after the demise of the agreement in 1989. He finds that his model performs quite well in projecting market events after the agreement, and also that a model describing the preferences of members of the ICA does indeed rationalize the outcomes observed under the international coffee market.

The research reported in this book has highlighted a number of factors behind the evolution of trade and investment in the food sector. Imperfect competition, product differentiation, and/or economies of scale were influencing firm strategies in each specific industry examined here. The new methods used have shown the power of recent advances in theory to explain behavior in these markets, and this research has shown that those methods can be implemented empirically. The research has also shown how policies will operate under conditions of imperfect competition, and that policy and market outcomes may differ under these circumstances.

Several chapters highlighted the need to tailor the theory and the empirical methods employed to specific market institutions under study. The diversity of problems encountered dictated the wide variety of methods used in this book, involving approaches taken from trade, marketing, and industrial organization approaches. Some approaches begin to bridge the gap in analytical methods of these disciplines, a step already begun through the introduction of industrial organization issues into the "New Trade Theory" and the consideration of trade in the Industrial Organization literature. This book represents a beginning in an important and potentially fruitful area of research for both trade and marketing economists in the future.

References

Abreu, D. and A. Sen. 1991. "Virtual Implementation in Nash Equilibrium." *Econometrica* 59:997-1021.
Bain, J.S. 1951. "Relation of Profit Rate to Industry Concentration: American Manufacturing, 1936-1940." *Quarterly Journal of Economics* 65:283-324.

Carlton, D.W. and J.M. Perloff. 1994. *Modern Industrial Organization, Second Edition*. New York, NY: Harper Collins College Publishers.

CIFAR. 1992. *Global Company Handbook*. Princeton, NJ: Center for Financial Analysis and Research.

Ethier, W.J. 1994. "Conceptual Foundations from Trade, Multinational Firms, and Foreign Direct Investment Theory," in M. Bredahl, P. Abbott, and M. Reed, eds., *Competitiveness in International Food Markets*. Boulder, CO: Westview Press.

Handy, C.R. and D.R. Henderson. 1994. "Foreign Direct Investment in Food Manufacturing Industries," in M. Bredahl, P. Abbott, and M. Reed, eds., *Competitiveness in International Food Markets*. Boulder, CO: Westview Press.

Helpman, E. and P. R. Krugman. 1985. *Market Structure and Foreign Trade: Increasing Returns, Imperfect Competition and the International Economy*. Cambridge, MA: MIT Press.

Jomini, P., J.F. Zeitsch, R. McDougall, A. Welsh, S. Brown, J. Hambley, and J. Kelly. 1991. "SALTER: A General Equilibrium Model of the World Economy." Canberra, Australia: Australian Industry Commission.

Knetter, M. 1989. "Price Discrimination by U.S. and German Exporters." *American Economic Review* 79:198-210.

_____. 1992. "Exchange Rates and Corporate Pricing Strategies." NBER Working Paper Series, No. 4151.

Krugman, P.R. 1987. "Pricing to Market When the Exchange Rate Changes," in S.W. Arndt and J.D. Richardson, eds., *Real-Financial Linkages Among Open Economies*. Cambridge, MA: MIT Press.

Lyons, B. 1981. "Price-Cost Margins, Market Structure and International Trade," in D. Currie, D. Peel, and W. Peters, eds., *Microeconomic Analysis: Essays in Microeconomics and Economic Development*. London: Croom Helm.

McDonald, S. and J.E. Lee. 1994. "Conceptual Foundations from Trade, Multinational Firms, and Foreign Direct Investment Theory," in M.E. Bredahl, P. Abbott, and M. Reed, eds., *Competitiveness in International Food Markets*. Boulder, CO: Westview Press.

Mundlak, Y. 1988. "Endogenous Technical Change and the Measurement of Productivity," in S.M. Capalbo and J.M. Antle, eds., *Agricultural Productivity Measurement and Explanation*. Washington, D.C.: Resources for the Future.

Pagoulatos, E. and R. Sorensen. 1976. "International Trade, International Investment, and Industrial Profitability of U.S. Manufacturing." *Southern Economic Journal* 42:425-434.

About the Editors and Contributors

Philip C. Abbott, professor, Department of Agricultural Economics, Purdue University.

Jean-Marc Bourgeon, researcher, Départment d'Economie, Institut National de la recherche Agronomique, France.

Maury E. Bredahl, professor, Department of Agricultural Economics, University of Missouri at Columbia.

Christos Cabolis, University of California, Santa Barbara.

Jay S. Coggins, assistant professor, Department of Agricultural Economics, University of Wisconsin at Madison.

John M. Connor, professor, Department of Agricultural Economics, Purdue University.

James R. Dayton, research associate, Department of Agricultural Economics, The Ohio State University.

Clarissa de Paz, former masters student, Department of Agricultural Economics and Business, University of Guelph.

Martha K. Field, Ph.D. candidate, Department of Agricultural Economics and Resource Economics, University of Connecticut.

Eric Giraud-Héraud, Station d'Economie et Sociologie Rurales, Institut National de la Recherche Agronomique, France.

Dennis R. Henderson, former branch chief, Economic Research Service, U.S. Department of Agriculture, Washington, D.C.

Thomas W. Hertel, professor, Department of Agricultural Economics, Purdue University.

Joseph G. Hirschberg, senior lecturer, Department of Economics, University of Melbourne, Australia.

Nicholas G. Kalaitzandonakes, assistant professor, Department of Agricultural Economics, University of Missouri at Columbia.

Yves Le Roux, researcher, Unite Politique Agricole et Modelisation, Station d'Economie et Sociologie Rurales, Institut National de la recherche Agronomique, France.

Rigoberto A. Lopez, associate professor, Department of Agricultural and Resource Economics, University of Connecticut.

Steve McCorriston, lecturer, Department of Agricultural Economics, University of Exeter, United Kingdom.

Yulin Ning, former graduate research assistant, Department of Agricultural Economics, University of Kentucky.

Emilio Pagoulatos, professor and head, Department of Agricultural and Resource Economics, University of Connecticut.

Timothy A. Park, assistant professor, Department of Agricultural Economics, University of Georgia.

Paul M. Patterson, research associate, Department of Agricultural Economics, Purdue University.

Daniel H. Pick, Economic Research Service, U.S. Department of Agriculture, Washington, D.C.

Alejandro Reca, research associate, Department of Agricultural Economics, Purdue University.

Michael R. Reed, professor, Department of Agricultural Economics, University of Kentucky.

Vincent Réquillart, Station d'Economie et Sociologie Rurales, Institut National de la recherche Agronomique, France.

Ian M. Sheldon, associate professor, Department of Agricultural Economics, The Ohio State University.

Erna van Duren, assistant professor, Department of Agricultural Economics and Business, University of Guelph.

Peter R. Vörös, former graduate research assistant, Department of Agricultural Economics, The Ohio State University.

Kai Wang, former graduate research assistant, Department of Agricultural Economics, Purdue University.

Printed and bound by CPI Group (UK) Ltd, Croydon, CR0 4YY

23/10/2024

01778232-0003